Jianhua Yan · Changming Du

Hydrogen Generation from Ethanol Using Plasma Reforming Technology

UNIVERSITY PRESS

大学出版社

Ⓐ Springer

图书在版编目（CIP）数据

等离子体重整乙醇制氢技术 =Hydrogen generation from ethanol using plasma reforming technology：英文 / 严建华，杜长明著 . —杭州：浙江大学出版社，2018.12

ISBN 978-7-308-18766-4

Ⅰ．① 等… Ⅱ．① 严… ② 杜… Ⅲ．① 等离子体—乙醇—制氢—英文 Ⅳ．①TE624.4

中国版本图书馆 CIP 数据核字 (2018) 第 275726 号

Hydrogen Generation from Ethanol Using Plasma Reforming Technology
严建华　杜长明　著

策　　划	许佳颖
责任编辑	金佩雯
责任校对	张凌静
封面设计	周　灵
出版发行	浙江大学出版社
	（杭州市天目山路 148 号　邮政编码 310007）
	（网址：http://www.zjupress.com）
排　　版	杭州中大图文设计有限公司
印　　刷	虎彩印艺股份有限公司
开　　本	710mm×1000mm　1/16
印　　张	7
字　　数	203 千
版 印 次	2018 年 12 月第 1 版　2018 年 12 月第 1 次印刷
书　　号	ISBN 978-7-308-18766-4
定　　价	98.00 元

版权所有　翻印必究　印装差错　负责调换

浙江大学出版社市场运营中心联系方式：0571-88925591；http://zjdxcbs.tall.com

Preface

Hydrogen has the potential to provide a clean and affordable energy supply that minimizes the dependence on oil, thereby enhancing the global economy and reducing environmental pollution. When used with traditional catalyst technology, onboard hydrogen production using hydrocarbons as starting materials is limited by the heavy weight of the device, a relatively long transient time and fouling, which increases the complexity of the onboard vehicle system. In situ hydrogen production from liquid feedstock using plasma or plasma-catalytic technology is an attractive alternative. In this book, hydrogen production from renewable resources such as ethanol by plasma or plasma-catalytic technologies is reviewed. These technologies have low NO_x emissions and low carbon footprints. Both the plasma and the plasma-catalytic systems have enormous potential for hydrogen production from renewable resources. Experimental studies have demonstrated the promising application of the combination of plasma and catalysts for hydrogen generation due to the synergistic effects. These technologies may offer a shortened learning curve and facilitate the entry of green reforming technologies onto the hydrogen market because of their reforming capacity and efficient hydrogen production. Relevant factors, including the input power, reactor geometry, temperature, carrier gas, and feedstock components, are discussed to better understand the alcohol reforming process using a non-thermal plasma reactor. Several models of alcohol reforming used to evaluate the reforming process are also reviewed. An overview of various plasma reactors and the efficiency of ethanol reforming are also discussed. The performances of the various systems are compared. The reforming efficiencies of different non-thermal arc plasma reactors are also discussed in terms of their characteristics and operating conditions. Finally, the directions of the research regarding the next generation of hydrogen production from fuel reforming are discussed. The future prospects of the plasma-catalytic approach for alcohol are exciting, and the synergistic effect of combining plasma and catalysts could play an essential role in the future. Authors gratefully acknowledge the work of Chao Shang, Zhiming Li, Dongwei Huang, Jianmin Mo, and Hongxia Li in the research.

Hangzhou, China Jianhua Yan
Guangzhou, China Changming Du

Contents

1 **Plasma for Ethanol Reforming** ... 001
   1.1 Hydrogen and Plasma ... 001
   1.2 Reforming Technologies of Liquid Fuel ... 003
      1.2.1 CO_2 Reforming ... 003
      1.2.2 Partial Oxidation Reforming ... 003
      1.2.3 Steam Reforming ... 004
      1.2.4 Autothermal Reforming ... 004
      1.2.5 Comparison Among Different Reforming Processes ... 004
   1.3 Hydrogen Production by Ethanol Reforming ... 006
      1.3.1 Catalytic Ethanol Reforming for Hydrogen Production ... 006
      1.3.2 Plasma Ethanol Reforming for Hydrogen Production ... 008
   References ... 010

2 **Non-thermal Arc Plasma for Ethanol Reforming and Hydrogen Production** ... 015
   2.1 Non-thermal Plasma ... 015
   2.2 Non-thermal Arc Plasma Reforming of Ethanol to Produce Hydrogen ... 017
   2.3 Factors Affecting on Plasma Reforming of Ethanol ... 019
      2.3.1 Effects of the Components of the Materials ... 019
      2.3.2 Effects of the Carrier Gas ... 020
      2.3.3 Effects of the Input Power ... 020
      2.3.4 Effects of Other Factors ... 021
   2.4 Comparison of the Non-thermal Plasma Reforming of Ethanol ... 022
   2.5 Developmental Trends of the Non-thermal Arc Plasma Reforming of Ethanol ... 025
   References ... 026

3 **Hydrogen from Ethanol by a Plasma Reforming System** ... 031
   3.1 Introduction ... 031
   3.2 Materials and Methods ... 031
      3.2.1 Experimental Setup ... 031
      3.2.2 Calculation ... 032
   3.3 Results and Discussion ... 034

3.3.1 Effect of the O/C Ratio 034
3.3.2 Effect of the S/C Ratio 037
3.3.3 Effect of the Input Power 038
3.3.4 Effect of the Ethanol Flow Rate 038
3.4 Conclusion 040
References 040

4 Hydrogen from Ethanol by a Miniaturized Plasma Reforming System 041
4.1 Introduction.. 041
4.2 Experimental Setup.. 041
4.3 Results and Discussion 043
4.3.1 Voltage-Current Characteristic 043
4.3.2 Effect of the O/C Ratio 044
4.3.3 Effect of the S/C Ratio 045
4.3.4 Effect of the Ethanol Flow Rate 046
4.4 Conclusion 048
References 048

5 Plasma-Catalytic Reforming for Hydrogen Generation from Ethanol 049
5.1 Introduction.. 049
5.2 Experimental Setup.. 049
5.2.1 Plasma-Catalytic Setup 049
5.2.2 Catalysts Characterization.. 051
5.3 Results and Discussion 053
5.3.1 Effect of the O/C Ratio 053
5.3.2 Effect of the S/C Ratio 055
5.4 Conclusion 057
References 057

6 Mechanism for the Plasma Reforming of Ethanol 059
6.1 Mechanism Analysis of the Single Plasma Reforming of Ethanol...... 059
6.1.1 Electron-Molecule Collision 059
6.1.2 Free Radical Reaction 067
6.1.3 The Generation and Conversion of the Main Products 074
6.1.4 Suppression and Removal of Carbon Deposition in the Reforming Process 077
6.1.5 Removal of NO_x in the Process of Reforming 080

- 6.2 Mechanism Analysis of the Plasma-Catalytic Reforming of Ethanol 081
  - 6.2.1 Related Mechanism of the Catalytic Reforming of Ethanol 082
  - 6.2.2 Effects of Plasma on the Surface Characteristics of the Catalyst 086
  - 6.2.3 The Surface Reaction of the Electronic/Radical-Catalyst 089
- 6.3 Comparison Between Plasma Reforming and Plasma-Catalyst Reforming 093
- 6.4 Summary 095
- References 097

7 Outlook 103

Index 104

Chapter 1
Plasma for Ethanol Reforming

1.1 Hydrogen and Plasma

With the growing global demand for energy, the depletion of fossil fuels as well as the awakening of public awareness of environmental protection, looking for an alternative energy source has attracted wide attention from the global energy industry. Significantly, hydrogen is an excellent alternative energy source. As a green energy source, hydrogen energy is characterized by the following advantages: (1) the only product of oxidation is water without any emission of pollution and green gas; (2) the combustion heat is up to 142 kJ/mol, which is much higher than that of conventional fuels such as gasoline, natural gas and coal; (3) hydrogen can be utilized as an efficient fuel for the proton exchange membrane fuel cell, which can be applied for vehicles and power plants, and (4) the addition of a certain percentage of hydrogen to the fuels can effectively increase the efficiency of the combustion of the engine. Hence, hydrogen is expected to become an important energy source in the future[1–4].

However, the direct utilization of hydrogen is still faced with many problems. The content of hydrogen in clean air is only 0.01%, and therefore the production of hydrogen by the separation of air is impossible. On the other hand, the density of hydrogen is very low and therefore the volumetric heating value is as low as 11 kJ/L; the liquefaction of hydrogen could be realized only when the temperature is as low as 20.3 K or under high pressure. What's more, the daily turnover of the liquefaction and the boiling loss during the recharging process are up 1%–2% and 10%–25%, respectively. The flammability of hydrogen also involves a hidden danger for its transportation. Therefore, finding an on-site hydrogen production process seems the best way to solve these problems[5–6].

The catalytic process is most widely used for hydrogen reforming with a significantly high conversion rate and hydrogen selectivity[7–8]. However, there are also many drawbacks to the catalytic process, such as the relatively higher cost, the coking and the inactivation resulting from pollution. Moreover, the catalytic process is generally operated under high temperature and the response time is always too long (several minutes are needed to start the system), which brings a challenge for the management of the heat. On the other hand, as for the application to vehicle batteries, the catalytic process is faced with the problem of the mass and volume which are too large in size[9]. Therefore non-thermal plasma reforming offers a new idea to avoid the problems discussed above.

Non-thermal plasma is also called non-equilibrium plasma, which is made up of energetic electrons, ions, radicals and the molecules and atoms in the ground state at a low temperature. Generally, non-thermal plasma is generated by gas discharge. By applying a strong electric field to a specific gas, electrically neutral gas can be ionized and charged species can be generated. Then the charged particles accelerate in the electric field. A large part of the energy is attained by lighter electrons, while the heavier species keeps a relatively lower temperature by colliding with the background gas. Therefore, the temperature of non-thermal plasma is up to 10,000–100,000 K (1–10 eV), while the background temperature is nearly the same as room temperature. Energetic electrons collide with neutral species, which leads to the formation of activated radicals, contributing to the high activation of plasma. Therefore, non-thermal plasma is characterized not only by an extremely high energy density, but also by the ability to initiate the reaction at a low temperature[10]. Nowadays, non-thermal plasma has been widely used in the reforming of energy, wherein the reforming of fuel for hydrogen production is attracting unprecedented attention. Plasma technologies used in reforming for hydrogen include non-thermal arc plasma[11–12], glow plasma[13], corona plasma[14], dielectric barrier discharge[15], microwave plasma[16], and so on.

Non-thermal plasma used in fuel reforming for hydrogen production has special advantages. Non-thermal plasma is driven by electricity, and has a short response time. What's more, the reforming efficiency can be modified through the adjustment of electrical parameters. There is a large amount of energetic electrons, radicals and excited molecules in plasma, so the excitation and ionization of the feedstock is promoted. Not all the molecules in the reaction zone are directly heated by discharge; the background temperature is relatively low, which leads to higher energy efficiency and lower energy consumption. Moreover, the corrosion of the electrodes can be weakened and therefore the construction and the materials of the set-up are more selective. Due to the higher energy density, the non-thermal plasma reforming reactor is characterized by a small volume and a light weight, and therefore it is more advantageous than other conventional reforming processes for application on vehicles. Moreover, such a reforming system can operate in ambient temperature with various carrier gases (such as N_2, O_2, Ar, He, air and steam). Thus, this reforming system is flexible and simple.

It is worth mentioning that the selectivity of the hydrogen production of plasma reforming still needs to be improved, including the optimation of the set-up construction, the size, the discharge ways as well as the coupling with the catalysis process to achieve a higher conversion efficiency and hydrogen production. Up to now, there have been many studies focusing on the plasma-catalysis process for methane reforming[17–19]. For other fuel reforming, however, relative literature is very limited.

Plasma-catalysis combined with a high activity of the plasma process and the high selectivity of the catalytic process. It can be predicted that the plasma-catalytic process will become a good choice in the field of hydrogen production, and that the optimation of the construction of the set-ups, the choice of an economically efficient catalyst may also become an interest of study.

1.2 Reforming Technologies of Liquid Fuel

The utilization of liquid fuel as reforming feedstock for vehicles and fuel barriers has obvious advantages such as (1) higher density, which leads to a higher hydrogen content per volume unit; (2) the state of liquid in ambient temperature and atm pressure make it easier to transport due to its flow ability and safety; (3) The source of liquid fuel is very widespread, which can be achieved by the fractionation of oil and the conversion of biomass. The most widely-used liquid fuel includes transportation fuels such as petrol, diesel, E85 and isooctane, alcohols such as methanol, ethanol and acetic acid and derivatives of hydrocarbons such as glycerol and diethyl ether. Nowadays, the mainstream technologies for the reforming of liquid fuel include reforming of CO_2 (dry reforming), partial oxidation reforming, steam reforming and autothermal reforming.

1.2.1 CO_2 Reforming

CO_2 reforming is generally used for the production of syngas (mainly H_2 and CO). Syngas plays an important role in the chemical industry, such as the production of light olefins with high value, long-chain hydrocarbons and oxygenates[17]. During the reforming process, H_2 and CO are produced by the oxidation of the fuel. And CO_2 can be reduced to CO. This reaction is generally an endothermic reaction; eternal heat is needed to maintain the reaction. CO_2 reforming of liquid fuel is most widely-used for the reforming of alcohols and ethers, the general formula of which is as follows:

$$C_nH_mO_p + (n-p)CO_2 \rightarrow (2n-p)CO + \frac{m}{2}H_2 \qquad (1.1)$$

1.2.2 Partial Oxidation Reforming

Taking liquid fuel and oxygen as the reforming feedstock, CO_2 and H_2O are produced by the total oxidation of fuel when the amount of oxygen is sufficient. While CO and H_2O are produced by the partial oxidation of fuel when the amount of oxygen is insufficient. Generally, this process consists of an endothermic reaction. Therefore, the total oxidation of part of the fuel is needed to offer the energy in demand. The general formulas of partial oxidation and total oxidation are shown as follows, respectively:

$$C_nH_mO_p + \left(\frac{n}{2} - \frac{p}{2}\right)O_2 \rightarrow nCO + \frac{m}{2}H_2 \qquad (1.2)$$

and

$$C_nH_mO_p + \left(n + \frac{m}{4} - \frac{p}{2}\right)O_2 \rightarrow nCO_2 + \frac{m}{2}H_2O \qquad (1.3)$$

1.2.3 Steam Reforming

Steam reforming is most widely used in the reforming process to produce hydrogen. Taking steam as oxidant, liquid fuel can be oxidized to CO_2 and H_2. On the other hand, an additional H_2 yield can also be offered by steam. Generally, this process is consists of an extreme endothermic reaction. The general formula is shown as follows:

$$C_nH_mO_p + (2n - p)H_2O \rightarrow nCO_2 + \left(2n - p + \frac{m}{2}\right)H_2 \qquad (1.4)$$

1.2.4 Autothermal Reforming

In order to improve the yield of hydrogen and lower or reduce the demand for eternal thermal source of heat, adding some air and steam during the reforming of liquid fuel can create the reaction in the thermal equilibrium. The general formula is shown as follows:

$$C_nH_mO_p + xO_2 + (2n - 2x - p)H_2O \rightarrow nCO_2 + \left(2n - 2x - p + \frac{m}{2}\right)H_2 \qquad (1.5)$$

1.2.5 Comparison Among Different Reforming Processes

Taking ethanol as an example, the four kinds of reforming processes are presented in Table 1.1. As seen in Table 1.1, a relatively high reforming temperature is a common feature of these four processes. The hydrogen yield of the same amount of ethanol follows as: steam reforming > autothermal reforming > partial oxidation reforming > dry reforming. Through the thermal dynamics of ethanol by means of the Gibbs free energy minimization method, it has been found that when the reaction temperature is 1200–1300 K and the CO_2/ethanol ratio is 1.2–1.3, the highest hydrogen yield per mol ethanol of 2.85 mol can be achieved[20]. Through partial oxidation reforming, steam reforming and autothermal reforming,

Table 1.1 Comparison of different reforming processes (DR, POR, SR and ATR) in terms of the performance of hydrogen generation from ethanol

Reforming process	Reforming material	Reforming temperature(K)	Hydrogen yield (mol_{H_2}/mol_{EtOH})	Characteristics	References
Dry reforming	C_2H_5OH/CO_2	550–700	1.5–1.9	Endothermic reaction and external heat source are needed, the main products are H_2 and CO, and high valued materials such as carbon nanofiber can be achieved. The CO selectivity is too high, which is not suitable for the proton exchange membrane fuel cell; the relative study is very rare	[22–24]
Partial oxidation reforming	C_2H_5OH/O_2	600–800	0.8	Exothermic reaction and no external heat source are needed, the response time is very short. However, the hydrogen selectivity is very low and the deposition of coke is large, which is not suitable for hydrogen production	[25–27]
Steam reforming	C_2H_5OH/H_2O	600–800	2.5–5.0	Endothermic reaction and external heat source are needed. The reaction rate is very slow and the volume of the setup is very large. The conversion rate and hydrogen selectivity are very high, and CO yield is relatively low. It seems to be most widely-used in the hydrogen production process	[28–30]
Autothermal reforming	$C_2H_5OH/O_2/H_2O$	500–600	2.0–3.8	Equilibrium reaction and no external heat source are needed, characterized by both thermal characteristics of steam reforming and high hydrogen selectivity, low CO selectivity of steam reforming process	[7–8, 31]

it has been found that the highest hydrogen yield per mol ethanol up to 4.85 mol can be achieved when T = 1000 K, S/C = 5 and O/C = 0.5 [21]. It has also been pointed out that, very different from other hydrogen productions, the coking problem and low hydrogen yield are the inevitable problem of partial oxidation reforming.

1.3 Hydrogen Production by Ethanol Reforming

Taking ethanol as the reforming material for hydrogen production, this method is characterized by the following advantages: (1) ethanol is much easier and safer for transportation, storage and direct utilization as a kind of liquid fuel; (2) the materials for ethanol are widely available, and it can be produced not only with biomass like sucrose, starch and oil, but also with agriculture and forestry waste; (3) the hydrogen content of ethanol is very high, and therefore much more hydrogen can be produced with the same weight of ethanol compared with other fuels; (4) the boiling point of ethanol is very low, and therefore gaseous ethanol can be easily achieved for reforming and hydrogen production; (5) ethanol is biodegradable and therefore the bio-toxicity is relatively low; (6) there are not any sulfur components, and therefore sulfur-containing pollutant gases are not a problem during the reforming procedure and (7) ethanol can be miscible with water in any ratio[27, 32-33]. It is worth mentioning that the yield of agricultural and forestry residues in China is the highest in the world and just the dry weight of the annual yield of crop straw reaches 500 million tons, so taking ethanol as the reforming material in China is feasible[34]. Nowadays, among all the ethanol reforming technologies, catalysis reforming has been a mature process in industry, and studies on low-temperature plasma for ethanol reforming has also attracted wide attention.

1.3.1 Catalytic Ethanol Reforming for Hydrogen Production

The catalysis process is the most perfect, developed and most widely-used reforming process for hydrogen production, which is always followed by high conversion efficiency and hydrogen selectivity. The catalytic metals used in the earlier steam reforming and autothermal reforming mainly refer to precious metals such as palladium, ruthenium, rhodium, iridium and platinum. Later non-precious metals such as nickel, cobalt and copper gradually attracted great attention and became a focus of research[35-36]. As shown in Table 1.2, comparisons of various catalytic metals for ethanol reforming are listed, and comparisons of various non-precious metals for ethanol reforming are listed as well.

Table 1.2 Comparison of the catalytic performances of various active metals in terms of ethanol reforming for hydrogen generation[38–40]

Catalytic metal	Advantages	Disadvantages
Pd, Ru, Rh, Lr, Pb	High activity and selectivity, excellent anti-coking feature	Lack of resource, large cost and high reaction temperature (600–800 °C)
Ni	High activity and selectivity, promoting the cleavage of C–C bond and reducing the production of liquid products, lower catalytic temperature and low cost	High selectivity of CH_4 and CO in products and a coking problem
Cu	Promoting the cleavage of C–H and O–H bond, used with Zn and Ni to improve the performance	The production of Ethylene and coking, sintering and by-products
Co	Promoting the cleavage of C–C bond, high activity in low temperatures, high selectivity of Ni, inhibition machination of CO and less production of CH_4	High cost and toxicity and problems caused by coking and inactivation

As shown in Tables 1.2 and 1.3, there are some drawbacks to catalytic processes, though high conversion efficiency and hydrogen selectivity can be achieved. As an example of these drawbacks, the cost of the catalyst is always very high and the catalyst is easily faced with inactivation caused by coking or pollution[37].

Table 1.3 Comparison of some results of ethanol catalytic reforming on various non-precious metal catalysts

Catalysts (supported metal/supporter)	Ethanol/steam/air	Temperature (°C)	Space velocity (h^{-1})	Conversion efficiency (%)	Hydrogen selectivity (%)	References
20%Ni/ γ-Al_2O_3	1/3/0	700	~10	~100	~189	[38]
18%Co/Al_2O_3	1/3/0	400	~15	99	126–140	[39]
10%Fe–Ni/ Al_2O_3	1/3/0.5	600	10,000	99.61	~115	[40]
2%Cu–14%Ni/ SiO_2	1/3.7/0	600	12.7	100.0	~106	[41]

Many researchers have been devoted to developing a catalyst with excellent resistance and catalytic properties. However, the catalytic process has to operate under high temperatures (500–800 °C). Moreover, the response time is very long for the system to start up, it takes several minutes. All these features put a great challenge on heat management. Furthermore, the catalytic process is faced with many problems, such as the size of the equipment that is too large, transition time that is too long when it comes to application for vehicles and families. As a newly-developed technology, a low-temperature reforming process offers a totally new way to solve the problems listed above.

1.3.2 *Plasma Ethanol Reforming for Hydrogen Production*

In recent decades, many researchers have studied the characteristics of various low temperatures and tested their reforming performance. Generally speaking, low temperature plasma is characterized by non-equilibrium properties. For non-thermal plasma, electron temperature is up to the order of 10^5 K, while the background temperature is always below 10^3 K, and it even gets close to room temperature. Low-temperature plasma reforming for hydrogen production is characterized by the following advantages: (1) there is a large amount of activated radicals, ions, energetic electrons, excited atoms and excited molecules in the plasma atmosphere and these activated species; (2) the energy density of plasma is extremely high, which makes the plasma reactor more compact; (3) the plasma process can be driven by electricity and therefore the response time and switching time are very short; (4) the plasma process can be driven by electricity and a large part of the energy in plasma is transferred to energetic electrons instead of heating the background gas, and therefore energy efficiency is significantly high; (5) H, OH and O radicals can be generated when H_2O is injected into the plasma gas, and therefore various types of hydrocarbons and their derivatives can be efficiently degraded. Based on the last advantages, the plasma reforming process is attracting more and more attention. Nowadays, plasma technologies have been applied in the reforming process, which includes glow discharge, corona discharge, microwave discharge, dielectric barrier discharge (DBD) as well as non-thermal discharge, which will be mentioned in the last section of this work. The plasma reforming process is a very promising process for hydrogen production. Various plasma technologies will be introduced shortly, pros and cons of which are also presented by examples.

Regarding glow discharge, the positive ions attack the cathode and avalanche and then a self-sustained discharge is formed. The discharge current of the glow discharge is in the order of several mA. For the atmosphere glow discharge, the discharge current is limited by introducing the negative feedback in the discharge current and voltage to avoid the transition from glow to arc discharge[42-43]. The glow discharge system was applied for the steam reforming of ethanol[44], the energy efficiency of hydrogen yield was 115 kJ/mol H_2. Although H_2 selectivity in gaseous products was up to 80%, the degradation of ethanol was incomplete and the major products were CH_3CHO, with a hydrogen yield rate of 20%.

Corona discharge is a kind of low-current discharge. Generally, the atmosphere corona discharge is generated by the local electrical breakdown of the gas gap in the non-uniform electric field. Hence, for the corona discharge, the size of at least one electrode is much smaller than the electrode gap. Common structures of corona discharge include line-area or line-column configuration[45]. DC and AC corona discharge was used in methanol reformation for hydrogen production. It was found that the hydrogen yield of AC corona discharge was much higher, the energy cost was lower than 0.02 Wh/cm^3 H_2. Pulsed corona discharge was utilized for the reforming of ethanol and E85[46]. Total conversion and hydrogen selectivity up to 127% can be achieved with a H_2O/ethanol of 29.2. The influence of the pressure and discharge gap on reforming is being investigated.

The characteristic of DBD is that the there is at least one electrode which is coated with a barrier layer. When ionization occurs in the local position, a large amount of charge accumulates on the surface of dielectric barrier in several nanoseconds. Applying a high AC voltage to electrodes, a microwave discharge is generated and distributed randomly in space and time[44]. An atmosphere pressure DBD reactor was used for ethanol reforming, a conversion efficiency of up to 100% and an energy efficiency of 12,000 kJ/mol H_2 were achieved[47]. The influence of DBD plasma reactor (11.2 kHz, 18 kV) filled with quartz beads was investigated on ethanol reforming[44]. In this study, besides H_2, CO and CO_2, CH_4, C_2H_4 and C_2H_6 were also produced. When quartz beads with a diameter of 2.0 mm were used, an ethanol flow ration of 75% and input power of 100 W were set, 45% hydrogen selectivity was achieved.

In the microwave plasma system, the microwave energy generated is collected by the microwave generator and transferred into the internal energy of gas molecules, therefore, plasma is formed through excitation and ionization. The wavelength used in microwave plasma is generally on the order of centimeter or decimeter. A 2.45-GHz microwave hurricane plasma in atmospheric pressure was used for the reforming of methanol, ethanol and propanol[48]. In the experiment, the ethanol conversion efficiency was larger than 99% and the hydrogen selectivity was up to 98.4%. A similar system was also used for the degradation of ethanol, and production of CO and CO_2 was found by the Optical Emission Spectrometry. Except for H_2, by-products such as C_2, OH and CH were detected[47].

Although non-thermal free arc plasma is certified as non-thermal plasma, it is characterized by some thermal features. Therefore, it has the advantages of both thermal plasma and non-thermal plasma. Compared with other thermal plasmas, non-thermal free arc can get a high electron density and a spouted flow rate[49].

Nowadays, non-thermal arc plasma has been widely used in energy and in the environmental industry, such as the abatement of VOCs[50-51], surface modification[52-53] as well as reforming for hydrogen production[54-56]. In the last section, recent research achievements on non-thermal arc plasma ethanol reforming for hydrogen production will be introduced.

References

[1] Huang DW. Design and application of miniaturized nonthermal arc plasma for hydrogen generation from ethanol reforming. Sun Yat-sen University. 2014.

[2] Mohanty P, Patel M, Pant KK. Hydrogen production from steam reforming of acetic acid over Cu–Zn supported calcium aluminate. Bioresour Technol. 2012; 123(3): 558–565.

[3] Alberico E, Sponholz P, Cordes C, Nielsen M, Drexler HJ, Baumann W, Junge H, Beller M. Selective hydrogen production from methanol with a defined iron pincer catalyst under mild conditions. Angew Chem Int Edit. 2013; 52(52): 14162–14166.

[4] Mori D, Hirose K. Recent challenges of hydrogen storage technologies for fuel cell vehicles. Int J Hydrogen Energy. 2009; 34(10): 4569–4574.

[5] Brown LF. A comparative study of fuels for on-board hydrogen production for fuel-cell-powered automobiles. Int J Hydrogen Energy. 2001; 26(4): 381–397.

[6] Huber GW, Shabaker JW, Dumesic JA. Raney Ni–Sn catalyst for H_2 production from biomass-derived hydrocarbons. Cheminform. 2003; 300(5628): 2075–2077.

[7] Kugai J, Subramani V, Song CS, Engelhard MH, Chin YH. Effects of nanocrystalline CeO_2 supports on the properties and performance of Ni–Rh bimetallic catalyst for oxidative steam reforming of ethanol. J Catal. 2006; 238(2): 430–440.

[8] Velu S, Suzuki K, Vijayaraj M, Barman S, Gopinath CS. In situ XPS investigations of $Cu_{1-x}Ni_xZnAl$-mixed metal oxide catalysts used in the oxidative steam reforming of bio-ethanol. Appl Catal B Environ. 2005; 55(4): 287–299.

[9] Song LJ, Li XH, Zheng TL. Onboard hydrogen production from partial oxidation of dimethyl ether by spark discharge plasma reforming. Int J Hydrogen Energy. 2008; 33(19): 5060–5065.

[10] Desmet T, Morent R, De Geyter N, Leys C, Schacht E, Dubruel P. Nonthermal plasma technology as a versatile strategy for polymeric biomaterials surface modification: a review. Biomacromolecules. 2009; 10(9): 2351–2378.

[11] Li XD, Zhang H, Yan SX, Yan JH, Du CM. Hydrogen production from partial oxidation of methane using an AC rotating gliding arc reactor. IEEE Trans Plasma Sci. 2013; 41(1): 126–132.

[12] Pornmai K, Jindanin A, Sekiguchi H, Chavadej S. Synthesis gas production from CO_2-containing natural gas by combined steam reforming and partial oxidation in an AC gliding arc discharge. Plasma Chem Plasma Process. 2012; 32(4): 723–742.

[13] Li DH, Li X, Bai MG, Tao XM, Shang SY, Dai XY, Yin YX. CO_2 reforming of CH_4 by atmospheric pressure glow discharge plasma: a high conversion ability. Int J Hydrogen Energy. 2009; 34(1): 308–313.

[14] Aleknaviciute I, Karayiannis TG, Collins MW, Xanthos C. Methane decomposition under a corona discharge to generate CO_x-free hydrogen. Energy. 2013; 59: 432–439.

[15] Martini LM, Dilecce G, Guella G, Maranzana A, Tonachini G, Tosi P. Oxidation of CH_4 by CO_2 in a dielectric barrier discharge. Chem Phys Lett. 2014; 593(2): 55–60.

[16] Kim TS, Song S, Chun KM, Lee SH. An experimental study of syn-gas production via microwave plasma reforming of methane, iso-octane and gasoline. Energy. 2010; 35(6): 2734–2743.

[17] Tu X, Whitehead JC. Plasma–catalytic dry reforming of methane in an atmospheric dielectric barrier discharge: Understanding the synergistic effect at low temperature. Appl Catal B Environ. 2012; 125(33): 439–448.

[18] Pham MH, Goujard V, Tatibouet JM, Batiot-Dupeyrat C. Activation of methane and carbon dioxide in a dielectric-barrier discharge-plasma reactor to produce hydrocarbons—influence of La_2O_3/gamma-Al_2O_3 catalyst. Catal Today. 2011; 171(1): 67–71.

[19] Long HL, Shang SY, Tao XM, Yin YP, Dai XY. CO_2 reforming of CH_4 by combination of cold plasma jet and Ni/gamma-Al_2O_3 catalyst. Int J Hydrogen Energy. 2008; 33(20): 5510–5515.

[20] Wang WJ, Wang YQ. Dry reforming of ethanol for hydrogen production: thermodynamic investigation. Int J Hydrogen Energy. 2009; 34(13): 5382–5389.

[21] Sun SH, Yan W, Sun PQ, Chen JW. Thermodynamic analysis of ethanol reforming for hydrogen production. Energy. 2012; 44(1): 911–924.

[22] Jankhah S, Abatzoglou N, Gitzhofer F. Thermal and catalytic dry reforming and cracking of ethanol for hydrogen and carbon nanofilaments' production. Int J Hydrogen Energy. 2008; 33 (18): 4769–4779.

[23] De Oliveira-Vigier K, Abatzoglou N, Gitzhofer F. Dry-reforming of ethanol in the presence of a 316 stainless steel catalyst. Can J Chem Eng. 2005; 83(6): 978–984.

[24] da Silva AM, de Souza KR, Jacobs G, Graham UM, Davis BH, Mattos LV, Noronha FB. Steam and CO_2 reforming of ethanol over Rh/CeO_2 catalyst. Appl Catal B Environ. 2011; 102(1–2): 94–109.

[25] Mattos LV, Noronha FB. Partial oxidation of ethanol on supported Pt catalysts. J Power Sources. 2005; 145(1): 10–15.

[26] Mattos LV, Noronha FB. Hydrogen production for fuel cell applications by ethanol partial oxidation on Pt/CeO_2 catalysts: the effect of the reaction conditions and reaction mechanism. J Catal. 2005; 233(2): 453–463.

[27] Klouz V, Fierro V, Denton P, Katz H, Lisse JP, Bouvot-Mauduit S, Mirodatos C. Ethanol reforming for hydrogen production in a hybrid electric vehicle: process optimisation. J Power Sources. 2002; 105(1): 26–34.

[28] Lopez E, Divins NJ, Anzola A, Schbib S, Borio D, Llorca J. Ethanol steam reforming for hydrogen generation over structured catalysts. Int J Hydrogen Energy. 2013; 38(11): 4418–4428.

[29] Han SJ, Bang Y, Yoo J, Kang KH, Song JH, Seo JG, Song IK. Hydrogen production by steam reforming of ethanol over mesoporous Ni–Al_2O_3–ZrO_2 aerogel catalyst. Int J Hydrogen Energy. 2013; 38(35): 15119–15127.

[30] Nichele V, Signoretto M, Pinna F, Menegazzo F, Rossetti I, Cruciani G, Cerrato G, Di Michele A. Ni/ZrO_2 catalysts in ethanol steam reforming: inhibition of coke

formation by CaO-doping. Appl Catal B Environ. 2014; 150–151(1641): 12–20.
[31] Huang LH, Liu Q, Chen RR, Hsu AT. Hydrogen production via auto-thermal reforming of bio-ethanol: The role of iron in layered double hydroxide-derived $Ni_{0.35}Mg_{2.65}AlO_{4.5\pm\sigma}$ catalysts. Appl Catal A Gen. 2011; 393(1–2): 302–308.
[32] Cavallaro S, Freni S. Ethanol steam reforming in a molten carbonate fuel cell. A preliminary kinetic investigation. Int J Hydrogen Energy. 1996; 21(6): 465–469.
[33] Tsyganov D, Bundaleska N, Tatarova E, Ferreira CM. Ethanol reforming into hydrogen-rich gas applying microwave 'tornado'-type plasma. Int J Hydrogen Energy. 2013; 38(34): 14512–14530.
[34] Yi Z, Zhongli P, Ruihong Z. Overview of biomass pretreatment for cellulosic ethanol production. Int J Agric Biol Eng. 2009; 2(3): 51–68.
[35] Ni M, Leung DYC, Leung MKH. A review on reforming bio-ethanol for hydrogen production. Int J Hydrogen Energy. 2007; 32(15): 3238–3247.
[36] Rossetti I, Lasso J, Finocchio E, Ramis G, Nichele V, Signoretto M, Di Michele A. TiO_2-supported catalysts for the steam reforming of ethanol. Appl Catal A Gen. 2014; 477(42–53): 42–53.
[37] Gallagher MJ, Geiger R, Polevich A, Rabinovich A, Gutsol A, Fridman A. On-board plasma-assisted conversion of heavy hydrocarbons into synthesis gas. Fuel. 2010; 89(6): 1187–1192.
[38] Fatsikostas AN, Verykios XE. Reaction network of steam reforming of ethanol over Ni-based catalysts. J Catal. 2004; 225(2): 439–452.
[39] Batista MS, Santos RKS, Assaf EM, Assaf JM, Ticianelli EA. High efficiency steam reforming of ethanol by cobalt-based catalysts. J Power Sources. 2004; 134(1): 27–32.
[40] Huang LH, Xie J, Chen RR, Chu D, Chu W, Hsu AT. Effect of iron on durability of nickel-based catalysts in auto-thermal reforming of ethanol for hydrogen production. Int J Hydrogen Energ. 2008; 33(24): 7448–7456.
[41] Vizcaino AJ, Carriero A, Calles JA. Hydrogen production by ethanol steam reforming over Cu–Ni supported catalysts. Int J Hydrogen Energy. 2007; 32(10–11): 1450–1461.
[42] Li X, Tao XM, Yin YX. An atmospheric-pressure glow-discharge plasma jet and its application. IEEE Trans Plasma Sci. 2009; 37(6): 759–763.
[43] Yang Y, Shi JJ, Harry JE, Proctor J, Garner CP, Kong MG. Multilayer plasma patterns in paralleled and coupled atmospheric glow discharges. IEEE Trans Plasma Sci. 2005; 33(2): 298–299.
[44] Yan ZC, Chen L, Wang HL. Hydrogen generation by glow discharge plasma electrolysis of ethanol solutions. J Phys D Appl Phys. 2008; 41(15): 1525–1528.
[45] Desmet T, Morent R, De Geyter N, Leys C, Schacht E, Dubruel P. Nonthermal plasma technology as a versatile strategy for polymeric biomaterials surface modification: a review. Biomacromolecules. 2009; 10(9): 2351–2378.
[46] Hoang TQ, Zhu XL, Lobban LL, Mallinson RG. Effects of gap and elevated pressure on ethanol reforming in a non-thermal plasma reactor. J Phys D Appl Phys. 2011; 44(27): 8295–8300.

[47] Jimenez M, Rincon R, Marinas A, Calzada MD. Hydrogen production from ethanol decomposition by a microwave plasma: influence of the plasma gas flow. Int J Hydrogen Energy. 2013; 38(21): 8708–8719.

[48] Tatarova E, Bundaleska N, Dias FM, Tsyganov D, Saavedra R, Ferreira CM. Hydrogen production from alcohol reforming in a microwave 'tornado'-type plasma. Plasma Sources Sci Technol. 2013; 22(6): 65001–65009.

[49] Yu L, Li XD, Tu X, Wang Y, Lu SY, Yan JH. Decomposition of naphthalene by dc gliding arc gas gischarge. J Phys Chem A. 2010; 114(1): 360–368.

[50] Du CM, Yan JH, Cheron B. Decomposition of toluene in a gliding arc discharge plasma reactor. Plasma Sources Sci Technol. 2007; 16(4): 791–797.

[51] Kusano Y, Norrman K, Drews J, Leipold F, Singh SV, Morgen P, Bardenshtein A, Krebs N. Gliding arc surface treatment of glass-fiber-reinforced polyester enhanced by ultrasonic irradiation. Surf Coat Technol. 2011; 205(2): S490–S494.

[52] Kusano Y, Teodoru S, Leipold F, Andersen TL, Sorensen BF, Rozlosnik N, Michelsen PK. Gliding arc discharge—application for adhesion improvement of fibre reinforced polyester composites. Surf Coat Technol. 2008; 202(22–23): 5579–5582.

[53] Wang BW, Sun QM, Lu YJ, Yang ML, Yan WJ. Steam reforming of dimethyl ether by gliding arc gas discharge plasma for hydrogen production. Chin J Chem Eng. 2014; 22(1): 104–112.

[54] Bo Z, Yan JH, Li XD, Chi Y, Cen KF. Plasma assisted dry methane reforming using gliding arc gas discharge: effect of feed gases proportion. Int J Hydrogen Energy. 2008; 33(20): 5545–5553.

[55] Yang YC, Lee BJ, Chun YN. Characteristics of methane reforming using gliding arc reactor. Energy. 2009; 34(34): 172–177.

[56] Sreethawong T, Thakonpatthanakun P, Chavadej S. Partial oxidation of methane with air for synthesis gas production in a multistage gliding arc discharge system. Int J Hydrogen Energy. 2007; 32(8): 1067–1079.

Chapter 2
Non-thermal Arc Plasma for Ethanol Reforming and Hydrogen Production

2.1 Non-thermal Plasma

Non-thermal plasma, also called non-equilibrium arc plasma, is a kind of low temperature plasma. Thermal arc plasma is usually driven by a high power supply and characterized by a high current (several amperes to tens of amperes) and a small low voltage (tens of voltage). It is a kind of strongly self-sustaining high-temperature discharge. Unlike thermal plasma, non-thermal arc plasma is usually driven by a low AC or DC power supply and characterized by a high voltage (several thousand voltages) and a small low current (tens to hundreds amperes). Non-thermal plasma can generally be divided into non-thermal fixed arc plasma and non-thermal free arc plasma, and non-thermal free arc plasma can be divided into knife-shaped electrodes non-thermal plasma, semi-free rotated arc discharge and free rotated arc discharge. In the last section, various non-thermal arc ethanol reforming systems are introduced.

Non-thermal fixed plasma can be generated by a pair of opposite tip electrodes or rod-shaped electrodes[1–2]. For the reforming of liquid fuel, the most used reforming process and oxidation-steam reforming with this kind of discharge. There are two ways to feed the materials: (1) the pair of electrodes are immersed in the liquid phase, and two flows of air are input along the axis of the pair of electrodes and then a stable countercurrent gas passage can be generated in the electrode gaps; (2) an air flow is introduced to the discharge zone from the center bore of an electrode[1–2]. Due to the fact that the direction of the air flow and that of the arc are the same, the position and the length is relatively fixed. This kind of discharge is also called non-thermal fixed arc discharge.

Knife-shaped gliding arc plasma is one of the simplest non-thermal free arc discharges, with the characteristics of high power and high non-equilibrium. Hence, this kind of discharge is widely used in the preparation of materials[3], treatment of pollutants[4–5], disinfection[6–7] and fuel reforming[8]. A pair of this kind of discharge reactors is made up of a pair of coplanar opposite knife-shaped metal sheets. When applying a high voltage, an arc is generated at the smallest gap between the electrodes. Then the arc slides towards the direction to which the electrodes gradually increase, and therefore the arc grows more and more curved. As the arc grows, when the energy loss is lower than the input energy, the gas temperature decreases while the temperature of the electrons is still very high. At this time, free arc thermal plasma transfers to non-thermal plasma. When the arc increases till the length is too large, the applied voltage continues to supply the energy needed for sustaining it, and then the conductivity of the plasma channel plummets, leading to the quenching of the plasma. At the same time, a new arc is generated in the smallest gap between the electrodes. The processes mentioned above circulate continually and discharge is sustained.

Five knife-shaped gliding arcs with different sizes were utilized for diethyl ether reforming[8]. It was found that the size of the discharge zone and the reforming efficiency increased with an effective electrode length and/or a slope radius of the downstream region of the electrodes. Some researchers have also used three sheets of knife-shaped electrodes or the simplified bent rod electrode for the treatment of methane and tetrafluoroethane[9–10].

Semi-free rotating arc discharge and free discharge both use a coaxial hollow external electrode and a center electrode, their external electrode is columnar, trumpet-shaped or zoom tube-shaped, while the center electrode is rod-shaped[11–13]. In the semi-free rotation arc discharge, the center electrode is positioned outside the external electrode. When both ends of the electrodes produce an electrical breakdown, one end of the arc root is fixed to the center electrode, the other end wanders on the external electrode under the effect of a gas flow and/or a magnetic field. In the free rotation arc discharge, the center electrode covers the entire external electrodes, so both ends of the arc can move freely on the two electrodes under the effect of an airflow and/or a magnetic field. In order to ensure the full connection and the reaction between the plasma and the feed stream in the discharge area, the tangential feed method is generally used to produce a rotating air flow. The above two types of discharge have been used in areas such as the degradation of toluene and NH_3, the reforming of methane to produce hydrogen and the synthesis of TiO_2 nanopowder and so on[14–18].

Zoom electrode freedom arc discharge plasma is a kind of special free rotation arc discharge. The reactor has the appearance of a long cylinder, and its electrode is made up of the coaxial center electrode and the zoom tube external electrode, the center electrode is a thin metal rod and the external electrode is a metal convergent-divergent nozzle. The center electrode covers the whole nozzle, so the narrowest point of the nozzle is also the point between the electrons that has the smallest gap. There are a number of tangential inlet holes on the top of the reactor to make the feed stream mix fully and form a cyclone in front of the discharge area. When the cyclone flows towards the downstream along the zoom, its flow rate increases rapidly while it maintains the rotation, and the air pressure decreases. Electrical breakdown occurs to this cyclone at the narrowest point of the nozzle after AC high voltage is applied to the electrodes, and then the arc rotaries and stretches with the cyclone downstream until it quenches.

2.2 Non-thermal Arc Plasma Reforming of Ethanol to Produce Hydrogen

A non-thermal fixed arc plasma was designed and the reforming effect of the ethanol solution was assessed through experiments and numerical simulation[19, 29]. In its discharge configuration, both of the electrodes were immersed in an ethanol solution and the two flows of air in opposite directions flowed along the two electrodes and the electrical breakdown and reforming occurred to the air flows and the solution in the gap of the electrode. The raw materials of the reforming were 50% ethanol solution and air, the flow rate of the air was 38 cm^3/s, the discharge current was 300 mA and the power was about 100 W. It was suggested that the content of the hydrogen in the gas produced reaches the highest level when the mole fraction of ethanol in the ethanol solution is 50%; at this time the conversion of ethanol and the yield of hydrogen are about 50% and 15% respectively. Concentrations of H_2 and CO are almost the same in the gas produced, and based on that, the author speculated that the temperature inside the reactor was about 355 K.

Another kind of non-thermal fixed arc plasma was used to reform the alcohols, bio-oil and wood[1]. The reactor was made up of a quartz envelope (the length was 400 mm, the internal diameter was 30 mm) and two graphite electrodes, the electrode gap was 10 mm. The apparent diameter of the arc column was 2 mm under the conditions of discharge. There was a through hole on the upper electrode axis, gasification of the ethanol solution happened before the ethanol solution entered the discharge region via the through hole, and the graphite electrodes were heated by the discharge in the reforming process.

One kind of non-thermal pulsed plasma sliding arc reactor with a spray nozzle was designed to explore the hydrogen generation rate and the energy efficiency when using methanol, ethanol and propanol as raw materials. A planar knife-shaped electrode configuration was used in the reactor. In the experiment, an alcohol solution was added into the reaction zone in the form of a pulse with Ar as a carrier so the temperature of the liquid in the reactor only increased by 1–2 °C. The production rate of hydrogen was 3.4–5.0 μmol/s when the discharge power was 0.45 W and 40% aerosol solution was added, of which the net volume of ethanol was 4–20 ml/min, and the energy consumption per hydrogen produced reached 90 kJ/mol_{H_2} at the maximum flow. It was also found in this study that the production rate of hydrogen would increase significantly with the increase in the flow rate of the feed stream and they speculated that this was because of the shorter residence time and quicker quenching and therefore the radical reaction which would result in the transformation if the target product was limited. In addition, they also used this type of plasma to process a stream of water droplets, and obtained H_2 and H_2O_2 effectively at the low power of 0.3–0.45 W[20].

A reformer of semi-free arc plasma was proposed to produce hydrogen by the reforming of liquid fuels such as gasoline, heptane, ethanol and E85 and so on[11, 21-23]. The flow rate of the ethanol solution was set to 0.25 g/s and the discharge power was 1200 W in the experiment (equivalent to 18% of the heating value of feedstock ethanol at a low temperature). The theoretical calculation and the result of the experiment both showed that the best reforming conditions are O/C = 0.5 and S/C = 0.5, at which the conversion of ethanol is 65%, the yield of hydrogen is 35% and the energy consumption of per hydrogen produced is 106–120 kJ/mol$_{H_2}$, while, when using E85 as the raw material, the above values become 90%, 65% and 60–100 kJ/mol$_{H_2}$ under the condition of O/C = 1.16, S/C = 0.2. In addition, there was no carbon deposition on the device during the two-hour continuous operation, which shows that the device has a good stability[11].

A kind of convergent-divergent tube free arc discharge reactor was designed to explore the volt-ampere characteristic of this discharge and the reforming effect of bio-ethanol as the focus[13]. The external electrode of this reactor was a copper convergent-divergent nozzle (the length was 106 mm, the narrowest inside diameter was 10 mm), the central electrode was made up of a fine stainless steel column (the length was 300 mm, the diameter was 5 mm), so the narrowest distance between the two electrodes was 2.5 mm. The power used was generated from the transformation of the 220 V AC through the AC leakage transformer (50 Hz, 10 kV). There were four tangential inlet ports at the top of the reactor, one of which was equipped with a gas-liquid nozzle to gasify the ethanol solution and form a high-speed vortex flow. The major reformate was synthesis gas (H_2 and CO), and other by-products such as C_2H_2, C_2H_4, C_2H_6, CO_2, CH_4 and so on were also generated. Besides, the emission of nitrogen oxide in the reforming process was very low (<10 ppm) (1 ppm=0.001‰). The experiment explored the effect of reforming of the bio-ethanol through reaction parameters such as the O/C ratio, the S/C ratio, the amount of ethanol and power and so on. A higher conversion rate of ethanol (90%) and a general yield of H_2 (40%) were obtained in the study and the best conditions for ethanol reforming were S/C = 2.0, and O/C = 1.4–1.6. When the amount of ethanol added was 0.15 g/s, the O/C ratio was 1.4 and the S/C ratio was 2.0, energy consumption required to produce one unit of hydrogen was 72.92 kJ/mol, reaching the minimum level. In the experiment, an oscilloscope and high-speed photography were also used to analyze the discharge characteristics of the converging-diverging electrode arc discharge reactor and it was found that an increase in the air flow will result in an increase in the instability of the change of the voltage and the current and the arc wire would appear split under the high airflow, which could improve the reforming efficiency of ethanol to a certain degree. The above findings indicate that this reforming process has an underestimated potential in the field of renewable energy.

2.3 Factors Affecting on Plasma Reforming of Ethanol

Various factors will affect the results of the reforming in the reactions of plasma in reforming ethanol. For example, the steam reforming reaction of water and ethanol will have different reaction processes when adding different proportions of water into the reaction of alcohol reforming. H and OH radicals generated by water in the plasma discharge will help the decomposition of ethanol. In addition, oxygen in the air can be ionized O radicals in the discharge when using air as the carrier gas. Next we will discuss the important factors that affect alcohol reforming.

2.3.1 *Effects of the Components of the Materials*

The components of the raw materials play an important role in the generation of free radicals, thus affecting the efficiency of the reforming and the distribution of the product. O–H bond energy of H_2O is 497.1 kJ/mol, while C–H bond energy of the methanol molecule is 95.18 kJ/mol, so the reaction activity of methanol in the plasma is higher than water[24]. Reactive components such as hydrogen atoms and hydrogen radicals and so on can be generated by the dehydrogenation reaction of H_2O molecules under the effect of the plasma. A DFT study on the corona discharge methanol reforming was conducted, and it was found that the decomposition of water can promote the generation of CH_2OH when it exists, which is because the OH radicals that are generated from the decomposition of water molecules can cause the oxidation decomposition of methanol[25]. The maximum yield of hydrogen can be produced when there is an appropriate amount of hydrogen in the raw materials; one of the reasons is that O is generated in the decomposition of water in the process of the partial oxidation reforming of alcohols, thereby increasing the production of hydrogen. In addition, H atoms produced during the decomposition of water can combine with each other and generate hydrogen. Take methanol for example: the amount of OH radicals will decrease with an increase in the concentration of methanol, causing the production of acid and CO_2. The relationship between the O/C ratio, the S/C ratio and the reforming efficiency was studied, and it was beneficial for the degradation of ethanol and the generation of H_2 and with an increase in the content of oxygen in the carrier gas, the conversion of ethanol showed a trend in which it first increased and then slowed down[26]. When the content of water increases, O radicals and H radicals generated from the water become more active, thereby promoting the ethanol molecule to participate in the collision reaction and its degradation significantly. When the content of oxygen increases, oxygen-free radicals generated by the oxygen can combine with the free radicals that are produced by the decomposition of ethanol quickly and transform into hydrocarbons, finally resulting in the increase in the conversion rate of ethanol. The experiments were conducted to study the effects on the reforming of different amounts of ethanol and water, and the results showed that the reduction of the amount of water will result in the reduction of the yield of hydrogen and carbon dioxide and an increase in the yield of CO[27].

2.3.2 *Effects of the Carrier Gas*

The carrier gas is closely linked with the types of the radicals that are generated in the alcohol reforming processes, so it can significantly affect the generation of gaseous products. Oxygen can degrade into oxygen free radicals with a high reactivity in the plasma discharge process to promote the oxidation decomposition of ethanol. Nitrogen in the air can also produce nitrogen oxides in the plasma discharge process, which has impacts on the environment. In the air plasma, oxygen and nitrogen will generate oxygen radicals and nitrogen radicals respectively under the impact of electrons. These two radicals can react and generate NO, NO can be converted to NO_2 under the role of oxygen free radicals, oxygen or ozone. Nitrogen needs 9.8 eV energy to dissociate up to N radicals and the energy that the oxygen needs to dissociate up to oxygen radicals is 6 eV, thus it will help to control the production of nitrogen oxides by controlling the reaction conditions, especially the discharge power. In addition, as a cheap industrial gas, argon gas is often used as a carrier gas in the plasma system. Argon will not affect the chemical reaction as the carrier gas, but it will affect the electron energy distribution equation[28]. Although argon will produce Ar in the plasma discharge process, the rate of the reaction between Ar* and ethanol is lower than that of the reaction between the ethanol molecule and the electron. In addition, the reaction between Ar* and ethanol requires a certain amount of Ar* so this reaction is negligible[29]. The experiment of the glow discharge plasma reforming of ethanol was conducted, and the production of the CO_2 can be reduced effectively by using argon as the carrier gas[29]. In addition, the flow of argon will also have an impact on the reaction system. A decomposition experiment of microwave plasma ethanol was conducted, and the impacts of the flow of argon on the reforming reaction were analysed[30]. It was found that as the flow of argon rises from 0.5 L/min to 1.5 L/min, the temperature of the gas drops from 3700 K to 2500 K. It was shown that the relationship between the yield of hydrogen and the carrier gas in the non-thermal plasma is as follows: Ar > N_2 > Air ≈ O_2[31-32]. The reasons are as follows: (1) compared with $N_2(a^1\Pi_g$, 8.59 eV), Ar($1\Pi_1^0$, 11.83 eV) can transmit more energy to the methanol so a higher yield of hydrogen can be obtained in Ar; (2) electric dipole transitions from N_2 ($X^1\Sigma_g^+$) to N_2 ($a^1\Pi_g$) are limited, so the excited reaction of N_2 is slightly weaker than that of Ar; (3) as a background gas, O_2 and the effects of the air on promoting the reaction are weaker, O_2 can react with hydrogen radicals produced by water or H_2 rapidly and produce water.

2.3.3 *Effects of the Input Power*

It was found that the conversion of methane increases with the increase in power, and it has nothing to do with the frequency or change of the voltage; for the latter it was found that the energy and density of the electron is only closely related to the voltage, so the energy and density of the electrons could not be affected by adjusting the frequency to change the power[33-34]. Our lab found that when the frequency is fixed, the size of the input power will affect the hydrogen production performance of the oxidation reforming of ethanol by affecting the conversion of ethanol and the yield of H_2. An increase in the plasma power can improve the density of high-energy electrons, thereby strengthening the excitation reaction caused by the electron collision and promoting the conversion of the alcohol. There is an important relation between power and the density of the plasma and almost all the energy is transferred to the electrons in the non-thermal plasma. Under the condition of plasma, a large number of active ingredients (high energy electrons, free radicals and so on) generated in the electron-molecule degradation promotes the decomposition of alcohols. The energy density of the discharge region, the average energy and the number of high-energy electrons will increase with the increase in the input power. And the increase in the high-energy electrons has a positive effect on the small molecules such as hydrogen and so on. The thickness of the plasma sheath (i.e., the volume of the plasma) increases with an increase in the discharge power; this indicates that the reaction channels of the high-energy electron in the discharge plasma increase and that the probability that the collisions of the high-energy electrons lead to degradation of the fuel molecule will increase with the increase in the voltage. However, the voltage cannot increase indefinitely, and the discharge electrode will melt when the voltage increases to a certain value, so the effect of the reforming cannot be increased infinitely. It was also observed that the increase in voltage usually helps to improve the conversion rate of the alcohol. In order to further improve the rate of production, we should study the reforming mechanism of alcohol in a profound way, so that we can better control the process of plasma reforming[35].

2.3.4 Effects of Other Factors

Temperature can be used to indicate the energy level obtained in the excitation reaction of the radicals caused by the involvement in the plasma of the heavy particles in the plasma reaction[31]. Radical chemistry reaction is closely linked to the temperature, while the electron collision reaction and the electron adsorption are not very closely linked to the temperature[36]. Various unwanted byproducts such as acetaldehyde, acetone and ethylene will be produced when the reaction temperature is low in practice. The conversion of ethanol and the yield of hydrogen will increase when increasing the reaction temperature of the gas mixture, and meanwhile, the selectivity of H_2, CO, CO_2 and CH_4 will improve and the selectivity of acetaldehyde, acetone and ethylene will decrease[37]. Since the mixed gas in the post-reaction zone is still at a relatively high temperature, the conversion reaction still occurs effectively after the discharge quenches[38]. It is believed that the effect of high temperature is caused by the increase in the rate of the electron—molecule collision reaction[38]. It was found that the H_2/CO ratio will increase and the concentration of the hydrocarbon will decrease with an increase in the temperature.

The residence time is one of the important factors that affects the reforming of ethanol and the production of hydrogen. The efficiency of the reforming increases with the increase in the residence time because the time of the reactions between the active ingredients and the molecules extends. When flow rate of the raw materials is very high, particles generated by the materials will leave a high electric field region in the short residence time, resulting in the particles leaving the discharge region before ionizing has been completed, thus the efficiency of the reforming reduces. However, if the residence time is too long, the reverse reaction will occur, which will also reduce the efficiency of the reforming.

Other factors, such as electric field, pressure, humidity, ozone concentration, the density of the electrons and the active ingredients, can also affect the effect of the fuel reforming. For example, since the electron collision caused the initial degradation process and generated the active ingredients, the density of the electrons is very important in the plasma reforming of alcohol[31]. Currently in order to obtain a more efficient reforming system, more and more factors are attracting the attention of researchers. Research conducted around other factors is very important and a lot of progress has been made.

2.4 Comparison of the Non-thermal Plasma Reforming of Ethanol

Table 2.1 lists the effects of all types of non-thermal plasma reforming of ethanol to produce hydrogen in the literature regarding the reforming conditions (discharge power, carrier gas, the amount of ethanol and the ratio of alcohol and water) and the reforming indicators (conversion rate of ethanol, selectivity of hydrogen, production rate of hydrogen and specific energy requirement (SER), etc.) and Fig. 2.1 also takes the production rate of the hydrogen and the energy consumption of per unit hydrogen production as indexes and intuitively reflects the differences on the hydrogen production performances of the ethanol reforming of different types of non-thermal plasma[1, 11, 13, 39-53]. To obtain a higher production rate of hydrogen during the reforming, the discharge device is required to have a larger amount of ethanol processing. The figure shows that the discharge types that have the largest amount of ethanol processing are glow discharge and non-thermal arc discharge, so these two types of discharge can obtain the maximum production rate of hydrogen. The discharge types that have the lowest energy consumption of per

Table 2.1 The effects of non-thermal plasma reforming of ethanol to produce hydrogen

Discharge type	Discharge power (W)	Carrier gas	Ethanol flux (g/s)	Ethanol-H_2O molar ratio	Conversion rate (%)	H_2 selectivity (%)	H_2 yield (%)	SER for H_2 (kJ/mol)	H_2 production rate (mol/h)	References
Glow discharge	850	Ar	0.51	1/0.013	87	22	20	115	26.6	[39]
Liquid phase glow discharge	72	–	0.027	1/0.39	22	100	22	186	1.39	[46]
	56	–	1.31×10⁻³	1/19	39	166	65	1008	0.2	[46]
Corona discharge	17	–	0.001	1/3.2	100	104	104	240	0.25	[40]
Corona discharge	~15	Air	1.32×10⁻³	1/0	90.1	74–85	~70	249	0.22	[53]
Pulsed corona discharge	~15	–	1.32×10⁻³	1/6	99	115	114	153	0.353	[47–48]
Pulsed corona discharge coupled with catalysis	~15	–	1.32	1/6	97	184	178	98	0.552	[48]
DBD discharge	100	Ar	0.053	1/0.85	–	–	45	64	5.6	[42]
DBD discharge	42	Air	–	1/0	30	105	32	–	–	[59]
DBD discharge	240	–	0.0247	1/1.08	65	94	61	244	3.54	[49]
DBD discharge	~30	–	2×10⁻³	1/1.28	65–100	56	36–56	1080	0.096	[41]
DBD discharge	40	Air	2.4×10⁻³	1/20	85–88	30–32	27	820	0.175	[50]
Surface microwave discharge	60	–	2.56×10⁻⁴	1/0	~100	~100	~100	3600	~0.06	[51]
	60	–	2.56×10⁻⁴	1/3	~100	~133	~133	2700	~0.08	
Surface microwave discharge	200	Ar	6.1×10⁻⁵	1/0	100	70–83	70–83	1300	0.55	[44]
Rotating microwave discharge	450	Ar	5×10⁻⁴	1/0	~99	96	95	13,500	0.12	[43]

(To be continued)

(Table 2.1)

Discharge type	Discharge power (W)	Carrier gas	Ethanol flux (g/s)	Ethanol-H_2O molar ratio	Conversion rate (%)	H_2 selectivity (%)	H_2 yield (%)	SER for H_2 (kJ/mol)	H_2 production rate (mol/h)	References
Rotating microwave discharge	450	Ar/Steam	5×10^{-4}	1/1.47	98.40	117	111	11,600	0.14	[52]
Non-thermal fixed arc discharge	100	Air	–	1/0	50	30	15	~100	–	[12, 19]
Liquid non-thermal arc plasma	300	Air	0.0256	1/2.56	100	62	62	291	3.7	[45]
Non-thermal fixed arc discharge	150–190	Air	0.03–0.14	1/0.82	49	104–129	51–64	~40	1.4–5.5	[1]
Pulsed dimensional knife-shaped glidarc	0.45	Ar	3×10^{-6}–15×10^{-6}	1/3.83	–	–	8–28	90	0.012–0.018	[19]
Semi free rotating arc discharge	1200	Air	0.25	1/0	65	~55	35	106–120	~40	[11]
Scaling tube free rotating arc discharge	100–300	Air	0.1–0.35	1/1.6–8.0	50–90	~40	~30	73	5.8	[13]
Micro semi-non-thermal fixed arc	11–13	Air	0.1	1/0.88	56	24	13	16	3.56	[60]
Micro semi-non-thermal fixed arc	12–14	Air	0.1	1/2	81	40	32	6.7	7.96	[60]

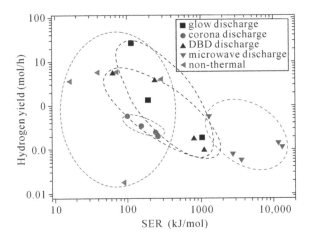

Fig. 2.1 Comparison of reforming behaviors of hydrogen generation out of ethanol with various types of non-thermal plasmas[1, 11, 13, 40–54]

unit hydrogen production are non-thermal arc discharge and DBD discharge, wherein the energy consumption of per unit hydrogen production of non-thermal arc discharge is concentrated in the 40–100 kJ/mol$_{H_2}$. Comprehensive analysis shows that the non-thermal arc is a very good hydrogen producer in ethanol reforming. Taking all of the analysts into consideration, the non-thermal arc is a very good ethanol reforming hydrogen production process.

2.5 Developmental Trends of the Non-thermal Arc Plasma Reforming of Ethanol

As described above, the non-thermal plasmas all have a large number of high-energy electrons, so they can react rapidly with the particles that need to be treated in the reaction zone. However, this also leads to a lower selectivity of plasma reactions. In recent years, many studies have been dedicated to using the catalyst-assisted plasma system to overcome the aforementioned shortcomings and improve the reforming or processing efficiency of the plasma system. Currently catalytic-assisted plasma systems have been widely used in the field of methane reforming[55–57]. In addition, the Cu/ZnO/Al$_2$O$_3$ catalyst was added into the DBD discharge region and synergies between the two in the steam reforming of methanol were explored[58]. It was found that discharge can reduce the temperature that the catalyst needs to play its activity. The synergy between them is believed to originate from the catalyst's absorption of the intermediates of the plasma reaction. The spark discharge plasma-catalytic system was used to degrade dimethyl ether to produce hydrogen and it was found that hydrogen production can be increased significantly at a high temperature

(50–700 °C) after adding the iron catalysts[59]. In addition, it was also found that the use of a catalyst can increase the discharge capacity. It is worth mentioning that the catalyst Pt/TiO_2 of the water gas shift and the filler layer $Pt-Re/TiO_2$ were placed just below the cathode electrode, so that the catalyst can fully utilize the heat generated during the plasma reforming[48]. By selecting the appropriate operating conditions, the content of CO that was produced from the degradation of ethanol (about 30%) was decreased to about 0.8% at the airspeed of 12,000 $cm^3/(g \cdot h)$, and the content of hydrogen in the gas production was up to about 73%, which means that after the removal of CO, the gas production can be used as hydrogen fuel cells. So in the future, combining the catalyst and the non-thermal arc to improve the load capacity and hydrogen production of the system can be considered, and the research should be focused on the exploration of the catalyst with low cost, long life and good performance. In addition, the reforming reactor needs to be further optimized in configuration and size, and to ensure the higher stability of the discharge and level of capacity, the density of the energy should be improved while reducing the size of the reactor; and the miniaturized reforming reactor or even the mini-reactor will bring more application fields, which will also bring more requirements and challenges for the plasma reforming reactor. Therefore, a combination of the catalyst, miniaturization, cost reduction and diversification and multifunctionality will be the trend of the future development of non-thermal plasma arc reforming.

Furthermore, because non-thermal plasma is very promising in fuel reforming, in the preparation of nanomaterial, in surface modification of material, in medical sterilization and in the treatment of environmental protection, it can be predicted that normalized and universal non-thermal plasma technology can be realized, paving the way for a low-cost, diversified and multifunctional non-thermal plasma process.

References

[1] Arabi K, Aubry O, Khacef A, Cormier JM. Syngas production by plasma treatments of alcohols, bio-oils and wood. J Phys: Conf Ser. 2012; 406(1): 1–22.

[2] Shchedrin A, Levko D, Ryabtsev A, Chernyak V, Yukhymenko V, Olszewski S, Naumov VV, Prysiazhnevych IV, Solomenko EV, Demchina VP, Kudryvtsev VS. Plasma kinetics in electrical discharge in mixture of air, water and ethanol vapors for hydrogen enriched syngas production. Probl At Sci Tech. 2008(4): 159–162.

[3] Djowe AT, Laminsi S, Njopwouo D, Acayanka E, Gaigneaux EM. Surface modification of smectite clay induced by non-thermal gliding arc plasma at atmospheric pressure. Plasma Chem Plasma Process. 2013; 33(4): 707–723.

[4] Du CM, Zhang LL, Wang J, Zhang CR, Li HX, Xiong Y. Degradation of acid orange 7 by gliding arc discharge plasma in combination with advanced fenton catalysis. Plasma Chem Plasma Process. 2010; 30(6): 855–871.

[5] Du CM, Xiong Y, Zhang LL, Wang J, Jia SG, Chan CY, Shi TH. Degradation of acid orange 7 solution by air-liquid gliding arc discharge in combination with TiO_2 catalyst. J Adv Oxid Technol. 2011; 14(1): 17–22.
[6] Wang J, Du CM, Zhang LL, Zhang ZL, Zhang CR, Xiong Y. Progress in research of gliding arc discharge plasma for sterilization. J Environ Eng. 2010; 28(6): 113–117.
[7] Wang J, Yang YJ, Du CM. Basic Research on the Inactivation of Bacterium by Plasma Generated by Gliding Arc Discharge. Guangdong Chemical Industry. 2013(13): 72–74.
[8] Wang BW, Sun QM, Lu YJ, Yang ML, Yan WJ. Steam reforming of dimethyl ether by gliding arc gas discharge plasma for hydrogen production. Chin J Chem Eng. 2014; 22(1): 104–112.
[9] Garduno-Aparicio M, Estrada-Martinez N, Pacheco-Sotelo J, Pacheco-Pacheco M, Garcia-Ramirez M, Valdivia-Barrientos R, Rivera-Rodriguez C, Gonzalez JJ. Three-Phase centrifuged gliding-arc discharge for CH_4 treatment. IEEE Trans Plasma Sci. 2011; 39(11): 2890–2891.
[10] Pacheco J, Garcia M, Pacheco M, Valdivia R, Rivera C, Garduno M. Degradation of tetrafluoroethane using three-phase gliding arc plasma. In: 14th Latin American Workshop on Plasma Physics, 2012; 370(370): 2014.
[11] Petitpas G, Gonzalez-Aguilar J, Darmon A, Fulcheri L. Ethanol and E85 reforming assisted by a non-thermal arc discharge. Energy Fuel. 2010; 24(4): 2607–2613.
[12] Bromberg L, Cohn D, Rabinovich A, Alexeev N, Samokhin N, Hadidi K. Onboard plasmatron hydrogen production for improved vehicles. Hydrogen. 2006.
[13] Du CM, Li HX, Zhang L, Wang J, Huang DW, Xiao MD, Cai JW, Chen YB, Yan HL, Xiong Y, Xiong Y. Hydrogen production by steam-oxidative reforming of bio-ethanol assisted by Laval nozzle arc discharge. Int J Hydrogen Energy. 2012; 37(10): 8318–8329.
[14] Li HX. Design, characteristics and application of non-thermal Laval nozzle arc plasma reactor for hydrogen production of bio-ethanol reforming. Sun Yat-sen University. 2012.
[15] Lee DH, Kim KT, Kang HS, Jo S, Song YH. Optimization of NH_3 decomposition by control of discharge mode in a rotating Arc. Plasma Chem Plasma Process. 2014; 34(1): 111–124.
[16] Liu SX, Li XS, Zhu XB, Zhao TL, Liu JL, Zhu AM. Gliding arc plasma synthesis of crystalline TiO_2 nanopowders with high photocatalytic activity. Plasma Chem Plasma Process. 2013; 33(5): 827–838.
[17] Lu SY, Sun XM, Li XD, Yan JH, Du CM. Decomposition of toluene in a rotating glidarc discharge reactor. IEEE Trans Plasma Sci. 2012; 40(9): 2151–2156.
[18] Zhou ZP. Reforming of Methane for Hydrogen Production via Non-equilibrium Plasma. University of Science and Technology of China. 2012.
[19] Burlica R, Shih KY, Hnatiuc B, Locke BR. Hydrogen generation by pulsed gliding arc discharge plasma with sprays of alcohol solutions. Ind Eng Chem Res. 2011; 50(15): 9466–9470.

[20] Burlica R, Shih KY, Locke BR. Formation of H_2 and H_2O_2 in a water-spray gliding arc nonthermal plasma reactor. Ind Eng Chem Res. 2013; 52(37): 13516.
[21] Petitpas G, Gonzalez-Aguilar J, Darmon A, Fulcheri L. Ethanol and E85 reforming assisted by a non-thermal arc discharge. Energy Fuel. 2010; 24(4): 2607–2613.
[22] Rollier JD, Petitpas G, Gonzalez-Aguilar J, Darmon A, Fulcheri L, Metkemeijer R. Thermodynamics and kinetics analysis of gasoline reforming assisted by arc discharge. Energy Fuel. 2008; 22(3): 1888–1893.
[23] Rollier JD, Gonzalez-Aguilar J, Petitpas G, Darmon A, Fulcheri L, Metkemeijer R. Experimental study on gasoline reforming assisted by nonthermal arc discharge. Energy Fuel. 2008; 22(1): 556–560.
[24] Yan ZC, Li C, Lin WH. Hydrogen generation by glow discharge plasma electrolysis of methanol solutions. Int J Hydrogen Energy. 2009; 34(1): 48–55.
[25] Liu XZ, Liu CJ, Eliasson B. Hydrogen production from methanol using corona discharges. Chin Chem Lett. 2003; 14(6): 631–633.
[26] Wang BW, Lu YJ, Zhang X, Hu SH. Hydrogen generation from steam reforming of ethanol in dielectric barrier discharge. J Nat Gas Chem. 2011; 20(2): 151–154.
[27] Aubry O, Met C, Khacef A, Cormier JM. On the use of a non-thermal plasma reactor for ethanol steam reforming. Chem Eng J. 2005; 106(3): 241–247.
[28] Levko D, Tsymbalyuk A. Ethanol reforming in non-equilibrium plasma of glow discharge. Plasma Phys Rep +. 2012.
[29] Levko DS, Tsymbalyuk AN, Shchedrin AI. Plasma kinetics of ethanol conversion in a glow discharge. Plasma Phys Rep. 2012; 38(11): 913–921.
[30] Jimenez M, Rincon R, Marinas A, Calzada MD. Hydrogen production from ethanol decomposition by a microwave plasma: Influence of the plasma gas flow. Int J Hydrogen Energy. 2013; 38(21): 8708–8719.
[31] Kabashima H, Einaga H, Futamura S. Hydrogen generation from water, methane, and methanol with nonthermal plasma. IEEE Trans Ind Appl. 2003; 39(2): 340–345.
[32] Kabashima H, Einaga H, Futamura S. Hydrogen generation from water, methane, and methanol with nonthermal plasma. IEEE Ind Appl Soc. 2001: 680–685.
[33] Batiot-Dupeyrat C, Goujard V, Tatibouet JM. Use of a non-thermal plasma for the production of synthesis gas from biogas. Appl Catal A-Gen. 2009; 353(2): 228–235.
[34] Liu CJ, Wang JX, Yu KL, Eliasson B, Xia Q, Xue BZ, Zhang YH. Floating double probe characteristics of non-thermal plasmas in the presence of zeolite. J Electrostat. 2002; 54(2): 149–158.
[35] Sarmiento B, Brey JJ, Viera IG, Gonzalez-Elipe AR, Cotrino J, Rico VJ. Hydrogen production by reforming of hydrocarbons and alcohols in a dielectric barrier discharge. J Power Sources. 2007; 169(1): 140–143.
[36] Harling AM, Demidyuk V, Fischer SJ, Whitehead JC. Plasma-catalysis destruction of aromatics for environmental clean-up: effect of temperature and configuration. Appl Catal B-Environ. 2008; 82(3–4): 180–189.
[37] Mattos LV, Jacobs G, Davis BH, Noronha FB. Production of hydrogen from ethanol: review of reaction mechanism and catalyst deactivation. Chem Rev. 2012; 112(7):

4094–4123.
[38] Tsymbalyuk AN, Levko DS, Chernyak VY, Martysh EV, Nedybalyuk OA, Solomenko EV. Influence of the gas mixture temperature on the efficiency of synthesis gas production from ethanol in a nonequilibrium plasma. Tech Phys. 2013; 58(8): 1138–1143.
[39] Yan ZC, Chen L, Wang HL. Hydrogen generation by glow discharge plasma electrolysis of ethanol solutions. J Phys D Appl Phys. 2008; 41(15): 1525–1528.
[40] Hoang TQ, Zhu XL, Lobban LL, Mallinson RG. Effects of gap and elevated pressure on ethanol reforming in a non-thermal plasma reactor. J Phys D Appl Phys. 2011; 44(27): 8295–8300.
[41] Sarmiento B, Brey JJ, Viera IG, Gonzalez-Elipe AR, Cotrino J, Rico VJ. Hydrogen production by reforming of hydrocarbons and alcohols in a dielectric barrier discharge. J Power Sources. 2007; 169(1): 140–143.
[42] Hu YP, Li GS, Yang YD, Gao XH, Lu ZH. Hydrogen generation from hydro-ethanol reforming by DBD-plasma. Int J Hydrogen Energy. 2012; 37(1): 1044–1047.
[43] Tatarova E, Bundaleska N, Dias FM, Tsyganov D, Saavedra R, Ferreira CM. Hydrogen production from alcohol reforming in a microwave 'ornado'-type plasma. Plasma Sources Sci T. 2013; 22(6): 65001–65009.
[44] Jimenez M, Rincon R, Marinas A, Calzada MD. Hydrogen production from ethanol decomposition by a microwave plasma: influence of the plasma gas flow. Int J Hydrogen Energy. 2013; 38(21): 8708–8719.
[45] Arabi K, Aubry O, Khacef A, Cormier JM. Syngas production by plasma treatments of alcohols, bio-oils and wood. J Phys: Conf Ser. 2012; 406(1): 1–22.
[46] Chernyak VY, Olszewski SV, Yukhymenko V, Solomenko EV, Prysiazhnevych IV, Naumov VV, Levko DS, Shchedrin AI, Ryabtsev AV. Plasma-Assisted reforming of ethanol in dynamic plasma-liquid system: Experiments and modeling. IEEE Trans Plasma Sci. 2008; 36(6): 2933–2939.
[47] Aubry O, Met C, Khacef A, Cormier JM. On the use of a non-thermal plasma reactor for ethanol steam reforming. Chem Eng J. 2005; 106(3): 241–247.
[48] Zhu XL, Hoang T, Lobban LL, Mallinson RG. Plasma steam reforming of E85 for hydrogen rich gas production. J Phys D Appl Phys. 2011; 44(27): 8295–8300.
[49] Zhu XL, Hoang T, Lobban LL, Mallinson RG. Low CO content hydrogen production from bio-ethanol using a combined plasma reforming-catalytic water gas shift reactor. Appl Catal B-Environ. 2010; 94(3–4): 311–317.
[50] Hu Y, Geshen L, Gao X, Yan L. Experiment study on hydro-ethanol reforming hydrogen production by non-thermal plasma. J Wuhan Univ Technol, Transport Sci Eng. 2009; 33(5): 928–931.
[51] Wang BW, Lu YJ, Zhang X, Hu SH. Hydrogen generation from steam reforming of ethanol in dielectric barrier discharge. J Nat Gas Chem. 2011; 20(2): 151–154.
[52] Yanguas-Gil A, Hueso JL, Cotrino J, Caballero A, Gonzalez-Elipe AR. Reforming of ethanol in a microwave surface-wave plasma discharge. Appl Phys Lett. 2004;

85(18): 4004-4006.
[53] Bundaleska N, Tsyganov D, Tatarova E, Dias FM, Ferreira CM. Steam reforming of ethanol into hydrogen-rich gas using microwave Ar/water "tornado"—Type plasma. Int J Hydrogen Energy. 2014; 39(11): 5663–5670.
[54] Zhu XL, Hoang T, Lobban LL, Mallinson RG. Partial oxidation of ethanol using a non-equilibrium plasma. Int J Hydrogen Energy. 2014; 39(17): 9047–9056.
[55] Nozaki T, Okazaki K. Non-thermal plasma catalysis of methane: Principles, energy efficiency, and applications. Catal Today. 2013; 211(211): 29–38.
[56] Tu X, Gallon HJ, Twigg MV, Gorry PA, Whitehead JC. Dry reforming of methane over a Ni/Al_2O_3 catalyst in a coaxial dielectric barrier discharge reactor. J Phys D Appl Phys. 2011; 44(27): 8295–8300.
[57] Kim T, Jo S, Song YH, Lee DH. Synergetic mechanism of methanol-steam reforming reaction in a catalytic reactor with electric discharges. Appl Energy. 2014; 113(1): 1692–1699.
[58] Song LJ, Li XH. Hydrogen production from partial oxidation of dimethyl ether by plasma-catalyst reforming. J Cent South Univ. 2013; 20(12): 3764–3769.
[59] Youping H, Bin G, Fubing Y, Geshen L. Plasmatron of H_2-rich gas generation from ethanol. Proceedings of ISES solar world congress 2007 solar energy and human settlement, 2007; 1–5: 2786–2789.
[60] Huang DW. Design and application of miniaturized nonthermal arc plasma for hydrogen generation from ethanol reforming. Sun Yat-sen University. 2014.

Chapter 3
Hydrogen from Ethanol by a Plasma Reforming System

3.1 Introduction

Gliding arc discharge is famous as a "warm" discharge with a temperature in the range of 2000–4000 K that can produce effective plasma with simultaneously high productivity and good selectivity. There are several disadvantages in the current gliding arc reactor with knife-shape or 2D-planar for its geometry, such as a low fuel conversion rate because of less contact time and collision frequency between feedstock and plasma[2]. The steam-oxidative reforming of bio-ethanol by a Laval nozzle arc discharge (LNAD) was studied at atmospheric pressure and room temperature without extra pre-heating. Unlike traditional gliding arc discharge, the gliding arc combines the advantages of both the 3-D cylindrical "tornado" type and the supersonic/subsonic discharge. Generally, the superiorities of this type of discharge are as follows:

(1) Any supply power and carrier gases can be employed to produce plasma;

(2) Plasmas generated by the carrier gas are characterized by the high density of electrons, high electric field and various reactive active species which accelerate the degradation reaction;

(3) A coverage of higher reaction volume and a good performance of the electrodes even under oxidizing conditions for the rotation of its arc root; a wide range of substances can be destructed and decomposed, facilitating its application in the industrial field;

(4) Few coke deposits on the inner wall of this novel reactor for the swirling flow, resulting in excellent plasma stability and security[1].

3.2 Materials and Methods

3.2.1 *Experimental Setup*

The experiment was conducted in a Laval nozzle arc discharge. The schematic view of the overall reactor is shown in Fig. 3.1. The plasma reactor was powered by a 10 kV AC power supply. An electronic watt-hour meter was used to monitor the input power applied to the reactor. A voltage regulator was employed to stabilize

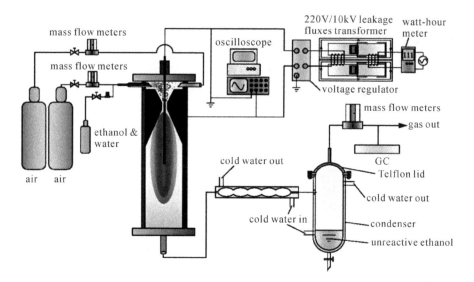

Fig. 3.1 The schematic view of the experimental setup

the voltage and current in the circuit which can be impacted by the operation of the plasma reactor. Moreover, the evolution of the voltage and the current in the plasma reactor under working conditions were determined by means of an oscilloscope with a high voltage probe (Tektronix P6035) and a current probe (Tektronix TP301A). A post-treatment gas system was positioned downstream for condensing the ethanol-water vapour and analyzing the exhaust gases as well as collecting the soot or coke generated in the process[1].

The plasma reactor was made up of a Laval nozzle made of copper with a length of 106 mm and a diameter of 10 mm, sealed by flanges at the top and at the bottom, which constituted the anode. The central cylindrical stainless rod with a length of 300 mm and a diameter of 5 mm on the axis of symmetry acted as the cathode. The minimum distance between two electrodes was 2.5 mm. The body of the reactor was made of stainless steel covered by a Teflon insulating layer, also of axial symmetry. The amounts of ethanol and water introduced into the air flow were controlled by peristaltic pumps at room temperature to maintain a constant inlet fuel flow rate. Note that this type of "tornado" movement was also preserved in the throat where the flow could be accelerated greatly and the distance between two electrodes was minimum[1].

3.2.2 Calculation

A mixture of water, air and ethanol can be produced in a controlled way by this system of device. The fuel and water mass flow rate are controlled in their liquid phase via two pumps in the range of 0.10–0.35 g/s and 0.06–0.47 g/s, respectively. The air flow rate is controlled, ranging from 139 to 1111 cm³/s correspondingly. The oxygen-to-carbon (O/C) ratio is a ratio of twice the molar oxygen flow versus the molar flow rate of the fuel carbon within an ethanol oxidation reaction scheme. Note that oxygen from CO_2 and H_2O is not considered in this ratio.

$$\frac{O}{C} \text{ ratio} = \frac{n_{\text{oxygen}}}{n_{\text{ethanol}}} = \frac{0.21 n_{\text{air}}}{n_{\text{ethanol}}} \quad (3.1)$$

Similar to the O/C ratio, the S/C (water-to-carbon) ratio is the ratio of the molar flow rate of H_2O to the molar flow rate of the fuel carbon. In this work, separate experiments with S/C = 0.8, 1.0, 1.2, 1.6, 2.0, 2.4, 3.2 and 4.0 were conducted to study their effects on the steam reforming of ethanol with no catalyst.

$$\frac{S}{C} \text{ ratio} = \frac{n_{H_2O}}{2 n_{\text{ethanol}}} \quad (3.2)$$

Ethanol conversion rate χ H_2 yield CO yield and energy efficiency η were employed to evaluate the steam reforming performance of ethanol, which can be expressed as follows:

$$\chi = \frac{(n_{\text{ethanol input}} - n_{\text{ethanol output}})}{n_{\text{ethanol input}}} \times 100\% \quad (3.3)$$

$$H_2 \text{ yield rate } (\%) = \frac{n_{H_2 \text{ in product}}}{3 n_{\text{ethanol input}}} \times 100\% \quad (3.4)$$

$$CO \text{ yield rate } (\%) = \frac{n_{CO \text{ in product}}}{2 n_{\text{ethanol input}}} \quad (3.5)$$

$$\eta = \frac{n_{H_2} \times LHV_{H_2} + n_{CO} \times LHV_{CO}}{n_{\text{ethanol}} \times LHV_{\text{ethanol}} + IPE} \times 100\% \quad (3.6)$$

LHV is defined as a lower heating value of each component. In addition, IPE is the abbreviation for input plasma energy. Energy efficiency is a significant parameter to estimate the conversion of ethanol in terms of energy. In addition, the specific energy input (SEI) for ethanol and the specific energy requirement (SER) of products (mainly H_2, CO, H_2 + CO) with the unit of kJ/mol were also calculated to evaluate the economic cost in industrial use.

$$\mathrm{SEI}_{(\text{ethanol})} = \frac{P(W) \times M_{\text{ethanol}}(\text{g/mol})}{[\text{ethanol}]_{\text{input}}(\text{g/s})} \times 10^{-3} \qquad (3.7)$$

$$\mathrm{SER}_{(H_2, CO, H_2 + CO)} = \frac{P(W) \times 0.001}{R[X](\text{mol/s})} \qquad (3.8)$$

where P is the input power, $R[X]$ is the production rate of the substance X (mol/s) and M_{ethanol} is the ethanol molecular mass (g/mol).

3.3 Results and Discussion

3.3.1 Effect of the O/C Ratio

To investigate the effect of O/C ratio on the steam reforming performances of ethanol, the volume flow rate of air was set between 139 and 1111 cm^3/s with fixed fuel flow rates (0.10, 0.15, 0.20 and 0.25 g/s respectively) to maintain the O/C ratio ranging from 0.6 to 2.0. Several reactions occurred in this experiment during the process of syngas production. First, for a better understanding of the oxygen mass on the steam-oxidative reforming of ethanol with the LNAD reactor, several experiments were performed to study the effects of the O/C ratio on the ethanol conversion rate. The relationship between the conversion rate and O/C ratio is expressed in Fig. 3.2. The maximum conversion rate of 90% ethanol was found when O/C = 1.6 with S/C = 2.0 and G_{ethanol} = 0.10 g/s. It can be explained that there were not enough O_2 molecules to break the ethanol molecule at a low O/C ratio, resulting in a weak conversion of ethanol as well as the production of syngas. As the O/C ratio increased, more heat was brought in, the reaction between ethanol and other species could be greatly promoted so that a higher conversion rate could be obtained. The main products are H_2 and CO when the reaction stays in the stage of partial oxidation which is a slight endothermic reaction[1].

The H_2 and CO yield reached the peak when the O/C ratio was around 1.5, which is higher than that of the reaction's stoichiometry. With the O/C ratio ranging from 0.8 to 1.5, the production rates of both CO and H_2 underwent acceleration. According to the calculation of the water mass detected in the condenser, we can conclude that the production of H_2O increases when the O/C ratio is greater than 1.5. This variation coincides basically with the change of H_2 yield, which implies that the composition of the main products shifts from H_2 to H_2O with the O/C ratio greater than 1.5. As the O/C grows from 1.0 to 1.6 with S/C = 2.0 and G_{ethanol} = 0.10 g/s, the production rates of H_2, CO and CO_2 rise from 1.29 to 1.90, 1.13 to 1.71 and 4.40×10^{-1} to 5.99×10^{-1} mmol/s, respectively. At the same time, the value of

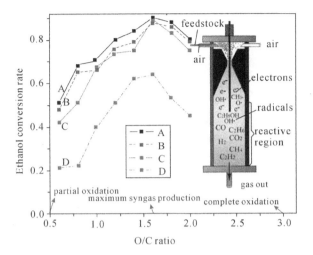

Fig. 3.2 Ethanol conversion rate versus O/C ratio with a different fuel flow rate and S/C ratio. (A) $G_{ethanol}$ = 0.10 g/s, S/C = 2.0; (B) $G_{ethanol}$ = 0.15 g/s, S/C = 2.0; (C) $G_{ethanol}$ = 0.10 g/s, S/C = 3; (D) $G_{ethanol}$ = 0.15 g/s, S/C = 3.2

CH_4 decreases from 5.11×10^{-1} to 4.06×10^{-1} mmol/s. When the oxidation of the mixture is sufficient to initiate the complete oxidation of ethanol, the addition of oxygen can also further contribute to the production of CO_2[1].

To investigate the influence of the O/C ratio on energy consumption, the term of specific energy requirement (SER) of different products (H_2, CO, H_2 + CO) was introduced, the definition of which was given in the previous section. The energy input with the unit of W as a function of the O/C ratio is expressed in Fig. 3.2. The input power supplied to the plasma grows with the increasing O/C ratio under a fixed ethanol flow rate. The mass of particles produced by the mixture of ethanol solution and air was faster at a higher O/C ratio with a fixed ethanol flow rate, resulting in an increase in the breakdown frequency, leading to an increase in the input power and the amount of arc filaments as well. However, the breakdown frequency was not infinite; it could be limited by the power. At the same time, it can also be observed that the input power experienced a decrease with a high speed flow rate of pure air (above 972 cm³/s), which can be explained by the fact that the particles generated from the mixture left the high electric field without ionization within a reduced residence time because of its fast speed, which decreased the resistance in the overall ionization region[1].

The SER of products as a function of various O/C ratios under given conditions is shown in Fig. 3.3. The minimal SERs of products, 72.92 kJ/mol (H_2), 80.20 kJ/mol (CO) and 38.19 kJ/mol (CO + H_2) respectively, were gained at O/C = 1.4 and S/C = 2.0 with an ethanol flow rate of 0.15 g/s. Unlike the case of S/C = 2.0, the SER of each product of S/C = 2.4 grows with an increase in the O/C ratio initially, reaching the maximum at O/C = 1.2 and the minimum at O/C = 1.6 respectively.

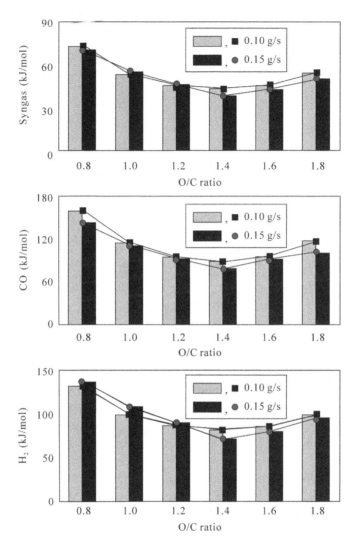

Fig. 3.3 Energy consumption dependence on O/C ratio with different fuel flow rates at S/C = 2.0

Note that there are several cases when the O/C ratio increases. First, as mentioned before, the input power ascends with an increase in the O/C ratio with a constant ethanol flow rate for the increasing number of arc filaments between electrodes and particles produced by the feedstock. Second, more energy should be supplied for the increasing conversion of ethanol and heat which increases together with the injection of oxygen as discussed before. Moreover, it is possible that further excitation or fragmentation of the products is formed with an increase in oxygen, which leads to further unnecessary energy consumption. When the acceleration of the power exceeds that of the production rate of the products, the SER will increase, and vice versa. Thus, has been demonstrated that the minimum SER occurs at O/C = 1.4 or 1.6 with S/C = 2.0 or 2.4, respectively. In a non-equilibrium plasma reactor, all of the energy is deposited in the electron components. The active species, produced in the electron-molecular processes, lead to chain reactions with the ethanol molecules. In the future, more studies should be done to further analyze the conversion mechanism of ethanol reforming assisted by plasmas, especially the main radical reactions converting ethanol to hydrogen[1].

3.3.2 Effect of the S/C Ratio

The S/C ratio is one of the most significant factors affecting the conversion of ethanol, especially the production of hydrogen. The effect of the S/C ratio on the production rates of H_2 and CO is expressed in Fig. 3.4. The production rate of H_2 was up to the maximum of about 1.90 mmol/s at S/C = 2.0 with the O/C ratio and the fuel flow rate of 1.6 and 0.10 g/s respectively, while the value of CO initially increased more slowly than hydrogen with the increasing S/C ratio. With the increases in the S/C ratio, the oxygen radical and hydroxyl radical from the water molecules became active enough to transfer the energy and collide with the ethanol molecules to promote its decomposition, causing the increase in syngas production expressed in Fig. 3.4. When the addition of water increased with the S/C larger than 2.0, the production rates of H_2 and CO decreased. One of the possible reasons may be that the additional water absorbs some heat released from the partial oxidation of ethanol, resulting in a decrease in the temperature in the reactor which has negative effects on the ethanol conversion[1].

The mole ratio of H_2 and CO reached their peak at the S/C ratio of 2.0–2.4, which indicates that the addition of water favors the steam reforming of ethanol and H_2 production. As the S/C ratio varied from 0.8 to 4.0 with other operating parameters remaining constant, the mole ratio of H_2 and CO increased up to a maximum of 1.14 at S/C = 2.0. Moreover, the mole ratio exceeded the unity with the S/C ratio in the range of 2.0–2.4 at $G_{ethanol}$ = 0.10 g/s. This can be explained by the concentration of free radicals as well as by the water-gas shift reaction, which had a favorable H_2 equilibrium under the conditions of large amounts of water added and low temperatures. Note that for temperatures lower than 1000 K in the investigated mixtures, the hydrogen atoms were generated mainly during the process of water dissociation by electron impact[1].

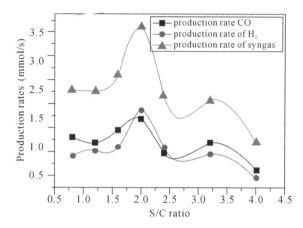

Fig. 3.4 The production rate of H_2 and CO as functions of the S/C ratio at O/C = 1.0 and $G_{ethanol}$ = 0.10 g/s

3.3.3 Effect of the Input Power

The input power is a key parameter affecting the performance of the reactor system; it can be changed by varying the applied voltage and pulsed frequency. This experiment was performed using a mixture of water, air and ethanol at O/C = 1.2, S/C = 2.0 and 3.2 respectively. From Fig. 3.5, it can be seen that the ethanol conversion rate was significantly affected by the input power. In addition, the H_2 yield slightly decreased with the growing discharge power. Future work on by-product analysis is necessary and is currently underway. The power has a close relationship with the density of plasma and almost all of the energy is deposited into the electron in the non-thermal plasma. A large amount of active species can be produced from the reaction of electrons and molecules, leading to the decomposition of ethanol[1].

3.3.4 Effect of the Ethanol Flow Rate

In the case of O/C = 1.0, the ethanol flow rate rose from 0.10 to 0.25 g/s, the measured H_2 and CO production rates grew from 1.29×10^{-3} to 2.56×10^{-3} mol/s and 1.13×10^{-3} to 2.40×10^{-3} mol/s, respectively. A similar variation of the production rate was obtained when O/C = 1.2, from which it can be concluded that more ethanol is decomposed within the unit time at a higher fuel flow rate, causing a higher H_2 or CO production rate. The detailed reasons may be that a larger amount of ethanol molecules, contained in the reforming process with a higher ethanol flow rate, results in an increase in the collision frequency between the ethanol molecules and the surrounding plasmas so as to raise the production rate of H_2 and CO.

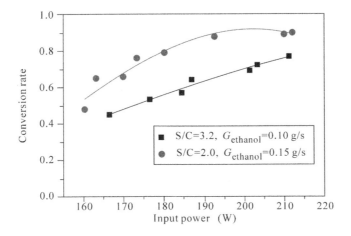

Fig. 3.5 The input power as a function of the ethanol conversion rate at different S/C ratios and ethanol flow rates

Dependence of energy consumption on the ethanol flow rate is shown in Fig. 3.6. The minimum SERs of H_2 and CO of 115.03 kJ/mol and 112.30 kJ/mol were achieved at the specific energy input of 36.8 kJ per ethanol mole with S/C = 1.6 and O/C = 1.2 at $G_{ethanol}$ = 0.25 g/s. This SEI corresponds to a power of approximately 200 W, 69% ethanol conversion and an air flow rate of 694 cm^3/s[1].

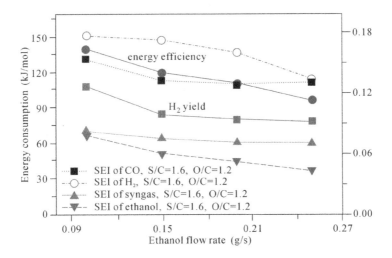

Fig. 3.6 The effect of the ethanol flow rate on the performance of ethanol steam-oxidative reforming

3.4 Conclusion

The steam-oxidative reforming of ethanol was studied using a Laval nozzle arc discharge (LNAD) for hydrogen production at the O/C ratio of 0.6–2.0, the S/C ratio of 0.8–4.0, a fuel flow rate of 0.10–0.30 g/s and atmospheric pressure. The mixture of ethanol and water was converted into the synthesis gas with C_2H_2, C_2H_4, CH_4 and CO_2. CO and H_2 were detected as the main compounds[1].

The LNAD reactor could also be used in some areas such as plasma ignition and flame control as well as hydrocarbon decomposition for the generation of carbon nano-tubes, fuel reforming and activation, used in the solar cell manufacturing process and so on.

References

[1] Du CM, Li HX, Zhang L, Wang J, Huang DW, Xiao MD, Cai JW, Chen YB, Yan HL, Xiong Y, Xiong Y. Hydrogen production by steam-oxidative reforming of bio-ethanol assisted by Laval nozzle arc discharge. Int J Hydrogen, Energ. 2012; 37(10): 8318–8329.

[2] Yan JH, Bo Z, Li XD, Chi Y, Cen KF. Plasma-assisted dry methane reforming using gliding arc gas discharge: effect of feed gases proportion. Int J Hydrogen Energy. 2008; 33(20): 5545–5553.

Chapter 4
Hydrogen from Ethanol by a Miniaturized Plasma Reforming System

4.1 Introduction

In the previous study, ethanol steam-oxidative reforming was performed with a Laval nozzle arc reactor, and the optimal operating conditions were found to be S/C = 2.0 and O/C = 1.4–1.6, with a minimum specific energy input in terms of ethanol of 55.44 kJ/mol and a minimum specific energy requirement of H_2 of 72.92 kJ/mol[2]. In the present study, a minimized arc plasmatron has been used to test the ethanol reforming at atmospheric pressure without additional heating, which has the advantages listed as follows[1]:

(1) A uniform distribution of active species results from the periodical rotation of arcs; elevated densities of energetic electrons and active species enable an excellent conversion extent and a faster reaction rate.

(2) Vaporize the liquid feed with high-speed air flows to get rid of the demand for an additional heat source for liquid vaporization.

(3) A special electrode configuration design, being able to minimize the erosion of the electrodes.

4.2 Experimental Setup

A minimized non-thermal arc plasma reactor was applied to convert bio-ethanol into hydrogen-rich gas by the steam-oxidative reforming method. The schematic diagram of the reactor is presented in Fig. 4.1. The non-thermal arc plasma reactor was mainly made up of a copper Laval nozzle (with an inner diameter of 2 mm at the throat) and a copper nail (with a diameter of 10 mm) with a pointed tip on one end. The body of the reactor was made of stainless steel. The copper nail located on the central axis acted as the central electrode, while the copper Laval acted as the other electrode. The minimum distance between these electrodes was set at about 0.27 mm, so the reactant was injected through a ring-shaped cross section (i.d. = 1.46 mm, o.d. = 2 mm) before entering the discharge zone as shown in the figure.

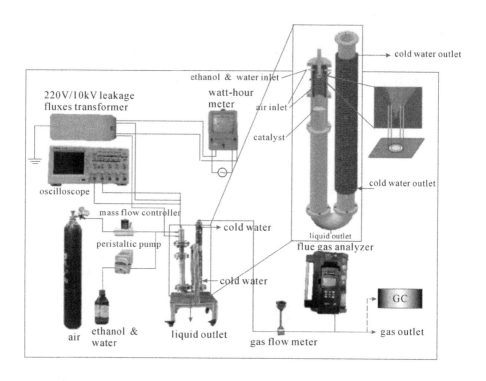

Fig. 4.1 The schematic diagram of a minimized non-thermal arc plasma reactor

The plasma reactor was driven by a 50 Hz high voltage AC power supply equipped with a 220 V/10 kV transformer. Here, employing the widely utilized AC power supply would make it possible to lower the costs of the device and of the operating, which would be beneficial for the popularization of this technology. A power meter was used to measure the power input consumed by the overall setup (mainly consumed by the transformer and the discharge reactor). The revolutions of the voltage and current in the discharge reactor when the plasma is on were analyzed with an oscilloscope (Tektronix 2024B) equipped with a high voltage probe (Tektronix P6035) and a current probe (Tektronix TP301A).

The reaction chamber inside the experimental reactor can be divided into three main parts: the mixing region, the discharge region and the post-discharge region. There were four orifices in various directions at the top of the reactor, with a gas-liquid spray nozzle inserted into one of them. The air stream was delivered through these orifices to form a vortex gas flow and the ethanol solution was introduced through the spray nozzle for a quick evaporation and a rapid mixing with air at room temperature. The flow rates of air and ethanol/water were maintained constant by a mass flow controller and a peristaltic pump, respectively. Then the atomized ethanol solutions entered the mixing region, where the electric breakdowns occurred and once a high voltage was applied between the electrodes, in the form of a high-speed whirl flow, then it flowed into the discharge region. Due to a sudden contraction at the narrowest position of the nozzle, the mixture stream might turn into a subsonic stream, or even a supersonic stream. The plasmas established in the discharge region consisted of various active species, such as high-energy electrons, highly-active radicals and molecules in their excited states, which then reacted with the mixture of ethanol and water, obtaining a variety of products such as H_2, CO, C_2H_2, C_2H_4, C_2H_6, CO_2, CH_4 and so forth. After the reactions, the reforming products were cooled with a condenser. To ensure the reliability of the obtained data, a flue gas analyzer was applied to monitor the concentration of H_2 and CO in the condensed gas products. Then the gas products were gathered after the concentrations of both H_2 and CO reached constant values, which indicated stable reactions and a constant production rate for the other main products (i.e. CO_2 and CH_4). The gathered exit gas was analyzed using a gas chromatography.

4.3 Results and Discussion

4.3.1 Voltage-Current Characteristic

The Voltage-Current (V-I) characteristic is one of the most important electrical characteristics of the non-thermal arc discharge. Because of its close link with the degree of ionization, the time-resolved variations of voltage and current can be used to characterize the evolution of the electric breakdowns and arc filaments.

The dynamical V-I characteristic during half a cycle of the discharge, which was obtained with the assistance of the oscilloscope, is plotted in Fig. 4.2. It was found that each half cycle experienced four phases according to the degree of ionization:

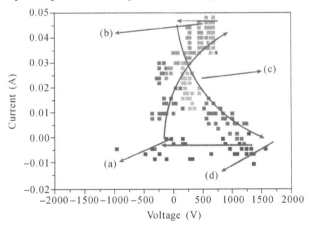

Fig. 4.2 The evolution of discharge current and voltage during half a discharge cycle with air flow rate of 0.48 m^3/h

(1) No ionization phase, where the voltage underwent a sudden elevation, causing an electric breakdown and the generation of the arc filaments when the voltage increased to a critical value, namely, the breakdown voltage;

(2) Full ionization phase; in this phase the current increased greatly while the voltage increased earlier, then dropped;

(3) High ionization phase; in this phase the voltage increased while the current decreased rapidly. The arc was simultaneously elongated by the air flow rate;

(4) Weak ionization phase, where the current decreased to 0 mA and the voltage dropped significantly. After being elongated to the maximum length due to the power input from the power supply, the arc column quenched as it could no longer afford the additional energy for maintaining a longer arc column.

4.3.2 Effect of the O/C Ratio

The influence of the O/C ratio within the non-thermal arc discharge reactor upon hydrogen reforming was tested by varying the air flow rate while holding the S/C ratio and the fuel flow rate constant, as can be seen in Fig. 4.3. As shown in Fig. 4.3, as the O/C ratio increased, the ethanol conversion augmented initially, reaching the peak value at the O/C of c.a. 0.50, and decreased afterwards. This is because inadequate oxygen gas participated in the oxidation of ethanol at lower O/C ratios, leading to a lower oxidizing extent and a relatively low ethanol conversion rate. As the O/C ratio grew, more oxygen molecules came into collision with ethanol molecules, so more ethanol was completely oxidized and more energy was released, which in turn enhanced the decomposition of ethanol. However, higher O/C ratios indicated larger air flow rates and shorter residence times of reactants within the reaction region, so more heat and unreacted ethanol were taken away, resulting in lower conversion extents.

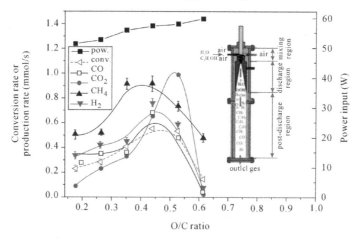

Fig. 4.3 Conversion rate (conv.), production rate and power input (pow.) as functions of the O/C ratio. Operating conditions: S/C = 0.43, ethanol flow rate = 0.10 g/s, without a catalyst

As the O/C ratio increased from 0.25 to 0.35, the production rate of CH_4 underwent a dramatic increase, while the production rates of H_2 and CO kept almost constant. Further increasing the O/C ratio to 0.45 would cause a quick acceleration in the generation of H_2 and CO. The maximum product yield of CH_4 occurred at O/C=0.40, but that of H_2 and CO occurred at O/C=0.45, which is slightly greater than the former one. Note that the conversion rate under the same condition reached its peak value at O/C=0.50. This could be explained by the fact that increasing the O/C ratio means that more CH_4 is oxidized into CO and CO_2 by oxygen; however, further addition of oxygen will convert H_2 and CO into H_2O and CO_2, thus lowering the H_2 yield.

4.3.3 Effect of the S/C Ratio

In order to analyze the effect of the S/C ratio on the performance of ethanol reforming, separate experiments were performed at different S/C ratios while holding other conditions constant. During the discharge process, increasing the S/C ratio at a constant ethanol flow rate meant delivering more steam into the reactor, so the composition of reforming products would be modified due to the variation in the conductivity of the mixture stream and the participation of more water molecules. The effect of the S/C ratio upon ethanol reforming performance is illustrated in Fig. 4.4 while setting a constant O/C ratio and ethanol flow rate (0.44 and 0.10 g/s, respectively). In the experiments, all the production rates for the main reforming products were found to be elevated with an increasing S/C ratio and reached their

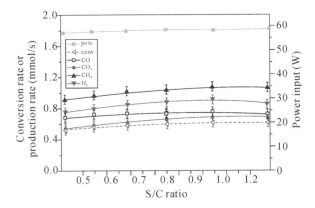

Fig. 4.4 Conversion rate (conv.), production rate and power input (pow.) as functions of the S/C ratio. Operating conditions: O/C = 0.44, ethanol flow rate = 0.10 g/s, respectively

peak values at an S/C ratio of 1.0, showing that adding water has a positive effect upon the conversion of ethanol to some extent. This is because as the S/C ratio increased, water molecules became easily motivated and dissociated due to the dissociation products out of oxygen and ethanol molecules as well as the energy released at the same time, which benefited the reforming reactions. Furthermore, if the S/C ratio increased from 0.43 to 1.05, H_2/CO and CO_2/CO ratios increased from 1.110 and 0.807 to 1.201 and 0.914, respectively, demonstrating that more CO reacts with water and is transformed into additional H_2 and CO_2 via the WGS reaction. On the other hand, the conversion rate decreased with increasing S/C ratios above 1.0 as a consequence of a decrease in the system temperature. At higher S/C ratios (above 1.0), adding more water did not mean more water molecules participating in the reactions. On the contrary, the introduction of too much water would decrease the energy deposited into each water molecule, which in turn resulted in a less effective reforming behavior.

4.3.4 *Effect of the Ethanol Flow Rate*

As one of the critical parameters of the operating conditions, the ethanol flow rate plays a significant role in ethanol reforming behaviors. Ethanol reforming performance is expressed in Fig. 4.5 as a function of the ethanol flow rate. In the experiments, the feeding rate of ethanol changed from 0.04 to 0.14 g/s with O/C = 0.44 and S/C = 0.43. The power input of the overall system was also an increasing function of the ethanol flow rate, and an increase in the ethanol flow rate from 0.04 to 0.14 g/s gradually elevated the power consumed from 50.9 to 60.4 W. As the ethanol flow rate increased, the conversion rate underwent an increase initially, and then declined, attaining a peak value of 55.4% at an ethanol flow rate of 0.10 g/s. The possible reason is as follows: under a given O/C ratio, higher ethanol flow rates indicate a faster air flow, leading to a quicker rotation of arc filaments and a good contact of the reactants with the plasma species, thus improving the conversion rates and enhancing the hydrogen yields. Increasing the ethanol flow rate above 0.10 g/s under a given O/C ratio, however, will reduce the conversion rates because of a shorter residence time during which more ethanol molecules leave the plasma zone without reaction. Moreover, an augmentation in the ethanol flow rate under a given S/C ratio may lead to a decrease in the reaction temperature because of the significant heat losses caused by the absorption of extra water. Also, an amplification of the ethanol flow rate may also reduce the energy density within the plasma zone, which limits the reactor temperature and the average electron energy, thus both the hydrogen production out of a unit ethanol mole and the energy efficiency fall. In terms of energy efficiency and hydrogen yield, the optimal ethanol flow rate is 0.10–0.12 g/s.

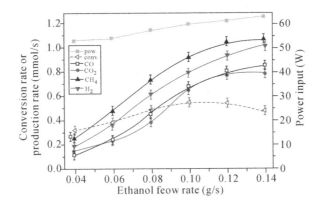

Fig. 4.5 Production rates, conversion rate (conv.) and power input (pow) as functions of the ethanol flow rate. Operating condition: O/C = 0.44, S/C = 0.43

The specific energy requirement (SER) of H_2 and syngas decrease with a growing ethanol flow rate. Though at higher ethanol flow rates, adding an ethanol-water mixture reduces the conversion rates, the production rates for the main products remain on the increase. Furthermore, the effect of the ethanol flow rate upon the power consumption is weak, so the SER of H_2 keeps on decreasing. For O/C = 0.44 and S/C = 0.43, the minimum SERs of H_2 and CO (59.3 and 70.6 kJ/mol, respectively) occur at an ethanol flow rate of 0.14 g/s, where the minimum specific energy input (SEI) of ethanol (19.8 kJ/mol) is achieved as well.

In the experiments, for O/C = 0.44, S/C = 1.28 and ethanol flow rate = 0.10 g/s, the power input of the plasma-catalytic reactor was 59.8 W, which meant that 27.8 kJ of electric energy would be consumed when 1 mol of ethanol flowed into the reactor, where 0.88 mol of ethanol was converted, producing 0.69 mol of H_2, 0.42 mol of CO, 0.82 mol of CH_4 and the other species. Taking H_2 and CO into consideration, the energy efficiency was no more than 22.4%. However, energy efficiency reached 73.9% when CH_4 was taken into account as well, so another plasma reforming reactor was needed to increase the selectivity of H_2. By setting appropriate O/C ratios, S/C ratio and power input, the latter reactor would further decompose a large amount of CH_4 produced from the former one into extra CO and H_2, Furthermore, the ethanol remaining unreacted in the former reactor would undergo conversion in the second reactor as well.

Because the energy efficiency of direct combustion is low, converting ethanol into hydrogen-rich gas for fuel cell application with optimum conditions and proper catalysts leads to higher energy efficiencies. Fuel cells can be widely used in many applications, such as power plants, transportation and telecommunications. Furthermore, the plasma-catalytic reforming does not demand a large quantity of electric energy. For example, in the present experiment, 0.69 mol of H_2 and 0.42 mol of CO, which contain a total energy content of 286 kJ, were obtained at the cost of 27.8 kJ of electric energy. The energy efficiency could be further increased if the second plasma reactor was applied, where more H_2 and CO would be formed from the unreacted ethanol and CH_4. In conclusion, ethanol reforming via a non-thermal arc plasma-catalytic system for fuel cell applications may be a promising solution for fossil fuel reservation and greenhouse effect prevention.

4.4 Conclusion

A non-thermal arc plasma reforming reactor was designed to study the oxidative-steam reforming of ethanol. The production of H_2 and CO was the main focus. In terms of energy efficiency, the optimal condition was determined to be O/C=0.5, S/C=1.0 and inlet ethanol=0.10 g/s. It has also been confirmed that the addition of a Ni/γ-Al_2O_3 catalyst contributes to an improved reforming performance with higher conversion rates and increased hydrogen production but little increase in the power consumed, and the SER for hydrogen decreases from 68.5 to 40.1 kJ/mol at O/C = 0.44, S/C=1.28 and inlet ethanol = 0.10 g/s. To further improve the reforming results, more attention must be paid to the modification of the structure and the size of plasma reforming reactors and electrodes, and the selection of catalysts with good catalytic behavior but less cost, which remains a significant task in the near future.

References

[1] Du CM, Li HX, Zhang L, Wang J, Huang DW, Xiao MD, Cai JW, Chen YB, Yan HL, Xiong Y, Xiong Y. Hydrogen production by steam-oxidative reforming of bio-ethanol assisted by Laval nozzle arc discharge. Int J Hydrogen, Energy. 2012; 37(10): 8318–8329.

[2] Du CM, Huang DW, Mo JM, Ma DY, Wang QK, Mo ZX, Ma SZ. Renewable hydrogen from ethanol by a miniaturized non-thermal arc plasma-catalytic reforming system. Int J Hydrogen Energy. 2014; 39: 9057–9069.

Chapter 5
Plasma-Catalytic Reforming for Hydrogen Generation from Ethanol

5.1 Introduction

Although benefits from energetic electrons contributed to the high activity and fast reaction rate in plasma reactions, the ethanol reforming assisted by non-thermal plasma usually reached undesirable product selectivity. Elevating the power input may increase the hydrogen selectivity to a certain extent, but it is not beneficial to improving the energy efficiency or to reducing the cost. A combination of plasma reforming with a catalyst seems to be a promising way to enhance the hydrogen yield and to noticeably reduce the energy required for hydrogen production simultaneously. Plasma-catalyst technologies have been sufficiently applied to methane reforming[2–4]; however, such applications on ethanol have seldom been seen in previous work. In this work, three non-precious metals (Ni, Cu and Co) have been selected to improve the reforming behaviors, and parametric experiments were performed to optimize the reforming conditions. It was found that the ethanol conversion rate and hydrogen selectivity were obviously enhanced, indicating that plasma-catalytic hybrid reforming is an attractive technique for fuel reforming.

5.2 Experimental Setup

5.2.1 Plasma-Catalytic Setup

The schematic diagram of the experimental setup is shown in Fig. 5.1(a). The non-thermal arc plasma reformer was made up of a copper converging-diverging nozzle and a copper nail, which acted as external and central electrodes, respectively. The minimum distance between the electrodes was set at 0.27 mm. The reformer was powered by a 10 kV transformer. The temporal variations of the discharge voltage and current were recorded by an oscilloscope. The catalyst pellets were filled in a stainless iron container located just under the external electrodes, so that the plasma could heat the catalyst, and a portion of the reactive plasma species (energetic electrons, radicals and molecule fragments) could reach the catalyst surface. Alumina-

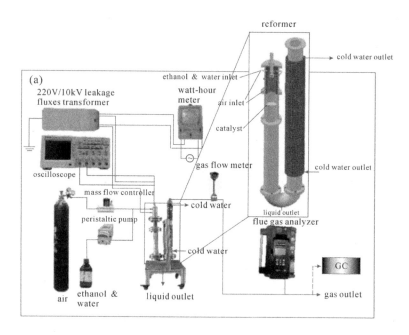
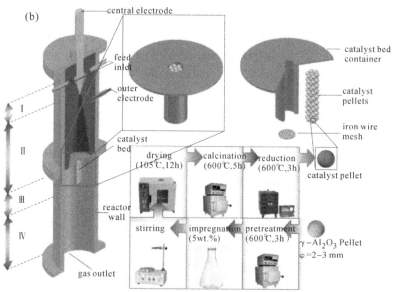

Fig. 5.1 Schematic diagrams of (a) the experimental setup and (b) structure of the non-thermal arc discharge reactor

supported non-precious metal catalysts (5% metal loading) were used[1]. In Fig. 5.1(b), the reaction chamber is divided into the following four main parts: I) the mixing region; II) the discharge region; III) the catalyst region and IV) the post-reaction region. The role of a gas–liquid spray nozzle and four tangential orifices formed the feedstock to a vortex stream, and then the vortex stream was introduced into the mixing region I); Due to the converging–diverging structure of the reformer, the vortex stream was accelerated dramatically in region II), and then it turned into a low-pressure rapid vortex flow within region III) as a consequence of the Laval nozzle effect. When applied with a continuous AC high voltage, electrical breakdowns occurred at the narrowest electrode gap, subsequently the discharge arc was pushed downstream by the low-pressure rapid vortex flow forming an extended plasma region, which exerted impacts on the catalyst surface. After the reforming reaction, the resulting products were cooled with a condenser prior to exit for gas analysis.

5.2.2 Catalysts Characterization

In Fig. 5.1(b), the catalysts (Ni/γ-Al_2O_3, Cu/γ-Al_2O_3, Co/γ-Al_2O_3) applied are prepared by the incipient wetness impregnation method. Granular γ-alumina with diameters of 2–3 mm and aqueous solutions of $Ni(NO_3)_2·6H_2O$, $Cu(NO_3)_2·3H_2O$ and $Co(NO_3)_2·6H_2O$ are used as support and metal precursors, respectively. Prior to impregnation, the support is calcined at 600 °C for 3 h. After impregnation, the samples are dried at 105 °C for 12 h, followed by the calcination at 600 °C for 5 h. Catalysts were reduced to 600 °C for 3 h in N_2/H_2 flow, and then were purged and cooled by N_2 stream till their temperature dropped to the ambient temperature.

The XRD (X-ray diffractometry) analysis revealed that after calcination, Ni^{2+} and Cu^{2+} interacted strongly with support and entered the structure of Al_2O_3, generating spinel phases; while the Co_3O_4 phase was formed in the case of the Co catalyst, showing that the interaction between Co^{2+} and the alumina support was weak. Furthermore, compared with the pattern near $2\theta=32°$, it is obvious that the diffraction peak of Co_3O_4 for Co catalyst is sharper than other patterns, which means that Co_3O_4 possesses a larger grain size and a poorer metal dispersion compared with that of $NiAl_2O_4$ and $CuAl_2O_4$. The crystal grain sizes of Co_3O_4, $NiAl_2O_4$ and $CuAl_2O_4$ are calculated as 20.2, 11.1 and 10.7 nm, respectively.

The support had the highest surface areas (326.5 m^2/g), which decreased remarkably after the addition of metal because the active metals covered the surface and filled the pores of the porous support. The surface areas decreased in the following order: Ni (224.7 m^2/g) > Cu (208.0 m^2/g) > Co (163.7 m^2/g). The grain sizes of the oxides of Ni and Cu were smaller due to the strong interaction with the support, thus the decreasing effect of the surface area, due to the filling of metal oxide, was less than that of the Co catalyst.

The SEM (scanning electron microscopy) images of the three metal catalysts, including both fresh and used (after 120 min operation), are shown in Fig. 5.2. It can be observed that the fine and uniform particles are prone to be formed on the catalyst surface after the plasma reaction, suggesting that plasma imposes effects on the physical properties of the catalyst surface, hence it gives rise to a uniform size and shape of the surface particles as well as an enhanced surface area. It can be seen from the figure that the particle sizes decrease in the following order: Ni < Cu < Co.

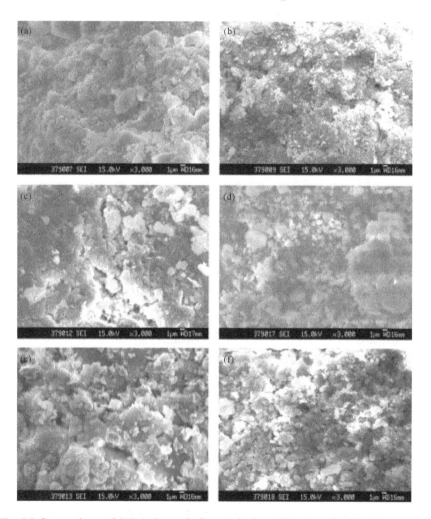

Fig. 5.2 Comparison of SEM photos before and after 120 min reforming in terms of different catalysts. (a) Fresh Ni/γ-Al$_2$O$_3$; (b) used Ni/γ-Al$_2$O$_3$; (c) fresh Co/γ-Al$_2$O$_3$; (d) used Co/γ-Al$_2$O$_3$; (e) fresh Cu/γ-Al$_2$O$_3$; (f) used Cu/γ-Al$_2$O$_3$

5.3 Results and Discussion

5.3.1 *Effect of the O/C Ratio*

The ethanol conversion rates as functions of the O/C ratio are shown in Fig. 5.3. For all the catalysts and the blank experiment, the conversion rates increased with an increasing O/C ratio in the range studied (0.3–0.6). With an O/C ratio increasing from 0.3 to 0.6, the reacted ethanol rose from 23.8% to 60.7%, while for Ni, Cu and Co catalysts, the corresponding values varied from 27.7%, 28.1% and 31.1% to 81.5%, 81.9% and 84.2%, respectively, demonstrating that the addition of catalysts favors the decomposition of the ethanol molecules with the order of Co > Ni > Cu. The best improvement was achieved by using the Co catalyst, whose conversion rate increased by 23.5%. It is worth noting that, for a plasma reformer without any packed bed, the converted ethanol achieved the maximum value at O/C=0.5, and a further increase of inlet oxygen led to a lowering of the conversion rate, while the highest conversion rates were obtained with O/C = 0.6 after a packed bed was added. Additionally, the energetic electrons and active species formed in plasma not only initiated elementary reactions in the discharge region and on the reactor wall, but they also reacted with the adsorbed molecules and molecular fragments on the active sites of the catalyst surface, leading to an improved decomposition rate of the ethanol[1].

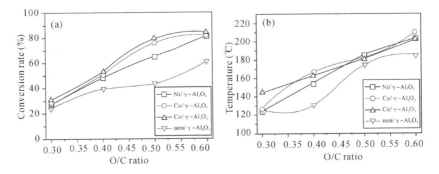

Fig. 5.3 Conversion rates (a) and reaction temperature (b) as functions of the O/C ratio

The reaction temperature was a significant parameter in catalyst reactions, which influenced the conversion rate and product distribution. The non-thermal arc plasma was powered by a high voltage supply device, providing rapid start-up and mode switching, and the reaction equilibrium could be well established in quite a short time (in 1–2 min). Based on this, a stable temperature could be easily reached in 2–3 min in the plasma-catalytic hybrid system. Furthermore, due to the non-equilibrium properties of non-thermal plasma, almost all the discharge power was transferred into energetic electrons instead of heating all of the molecules. When no external heat source was applied, the final temperature of the reactor was maintained by the heat released during the ethanol decomposition, compensating the heat loss due to the heat exchange and product transportation. As seen in Fig. 5.3(b), the reaction temperature rose with an increasing O/C ratio in a range of 120–210 °C, which is much lower than that in conventional catalytic processes (500–800 °C). At a lower O/C ratio (0.3), less oxygen was delivered into the discharge region, which led to lower concentrations of the active species as well as an incomplete mixing and interaction with the reactants, thus limiting the ethanol decomposition and the heat release. At a lower O/C ratio (0.3), the conversion rate and reaction temperature with Co catalyst was higher than others, which can be attributed to the high activity of Co at lower temperatures. In addition, a lower temperature did not favor the catalyst activity; therefore slight improvements in the conversion rate and reaction temperature were observed at a lower O/C ratio. However, further increasing the O/C ratio would counteract this advantage at higher temperatures. In general, a conclusion could be drawn: supplying more oxygen intensifies both the discharge-induced reactions and catalyst activities, obtaining a better reforming performance[1].

For the case of alumina support, the highest H_2 selectivity of 34.7% reached at O/C = 0.5, while the H_2 selectivities of different catalysts at identical O/C ratio were 40.2%, 37.2% and 37.9% for Ni, Cu and Co catalysts, respectively. Within the engaged O/C ratio, the highest H_2 selectivity (excess 40%) and the lowest CH_4 selectivity (not more than 40%) were achieved with the Ni catalyst, while the Co catalyst seemed to show no improvement in product selectivity when compared with the case using alumina support. For the Ni catalyst, the maximum H_2 selectivity of 46.3% was obtained at a lower O/C ratio (0.3), this value dropped to 39.6% at O/C = 0.6, while the selectivity for CO increased slightly with the enhancement of the O/C ratio. The case of Cu shares a similar tendency, with a greater increase in the selectivity of CO compared with Ni. With an increase of the O_2 concentration, more H_2 and CH_4 were oxidized into H_2O and CO. For the Co catalyst, unlike the former catalysts, the selectivities of H_2 and CO reached their peak values at O/C = 0.5, whereas the lowest CO_2 selectivity was exhibited. For all the catalysts, the selectivity for CH_4 decreased with an increase in the O/C ratio. When the O/C ratio increased from 0.3 to 0.5, CH_4 in product gas was oxidized into additional H_2 and CO, and a further increase in the supplied oxygen led to the oxidation of H_2 and CO into H_2O and CO_2, respectively. In terms of H_2 selectivity, the optimal O/C ratio for Ni and Cu was 0.3, while that of Co was 0.5[1].

As shown in Fig. 5.4, for Ni and Cu catalysts, the production rates of hydrogen increased with an increase in the O/C ratio, and the maximal values of 2.21 mmol/s (Ni) and 2.01 mmol/s (Cu) were achieved at O/C = 0.6; the maximal values for Co (2.06 mmol/s) and support (1.04 mmol/s) were achieved at O/C = 0.5, indicating that the hydrogen productivities are doubled due to the additions of catalysts. Hence, the additions of catalysts greatly enhanced the conversion efficiency and the hydrogen selectivity simultaneously. The SER is the ratio between the discharge power and the production rate. The discharge power is insensitive as a function of O/C ratios, which increased slightly with an increase in the O/C ratio (with a variation range of 13.31–14.72 W for catalysts), thereby the minimum and maximum SER for H_2 occurred at similar O/C ratios. The minimum SER for H_2 is not more than 6.7 kJ/mol which was reached at O/C = 0.6, S/C = 1.0 with an ethanol inlet of 0.10 g/s and with the Ni catalyst[1].

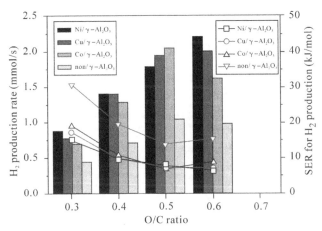

Fig. 5.4 Hydrogen production rates and relevant SER with different catalytic beds in plasma-catalytic reforming

5.3.2 *Effect of the S/C Ratio*

The conversion rates were studied as functions of the S/C ratio in Fig. 5.5(a), which revealed that the presence of catalysts favors higher conversion rates. For a moderate S/C ratio (1.0–1.5), the conversion performances of ethanol were in the order of Cu > Co > Ni, and the maximum conversion rate of 76.3% was obtained at O/C = 0.5 and S/C = 1.0 with the Cu catalyst. Increasing the steam injected, the optimal conversion rates were obtained at a moderate S/C ratio (1.0) with the Ni and Co catalysts; however, a lower S/C ratio (0.5), which is much lower than the stoichiometric ratio of the auto-thermal reformer, led to the highest conversion rate with the Ni catalyst. For all applied catalysts, the improved conversion rates compared with the case with a support decrease as the S/C ratio rose, showing that the increase in the S/C ratio was slightly beneficial to the synergetic effect between plasma and catalyst within reforming processes[1].

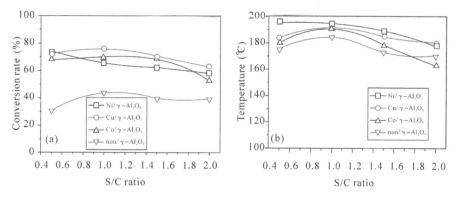

Fig. 5.5 Conversion rates (a) and reaction temperature (b) as functions of the S/C ratio

As shown in Fig. 5.5(b), the variations of the reaction temperature verses S/C ratio seemed to have a similar tendency with those of the conversion rates in a range of 160–200 °C. For the Cu and Co catalysts, the temperature increased with an increase in the S/C ratio until the optimal S/C ratio (1.0) was reached, and then it decreased. For the Ni catalyst, the reaction temperature dropped with an increase in the S/C ratio, with the highest value of 198 °C at S/C = 0.5. For a higher S/C ratio (>1.0), an addition of steam lowered the specific energy for each water molecule and slowed down the reaction rates. Furthermore, an elevation in steam flux means a short residence time, which leads to more unreacted ethanol and less heat released. It was found that the reaction temperature for support without catalysts was usually below that with catalysts, which could be explained by less decomposed ethanol[1].

For the case where catalysts were applied, the selectivity of CO decreased with the growing of the S/C ratio and the selectivity of CO_2 showed the opposite tendency. Due to the WGS reaction at a lower temperature (lower than 210 °C in our work), CO was converted into CO_2 by reacting with H_2O, resulting in an additional generation of H_2. Because more water leads to lowering down the reaction temperature, it limits the productivity and selectivity of H_2. Increasing the S/C ratio favored the steam reforming of CH_4, while a drop in temperature favored CH_4 generation out of ethanol incompletely decomposed. Therefore the selectivity of CH_4 seemed to be insensitive as functions for all catalysts. For the case with support, the selectivity of CH_4 increased with higher S/C ratios (>1.0). Compared with the control group, the presence of the Ni or Cu catalyst significantly reduced the CH_4 content in products with selectivity below 40%, while the presence of Ni is more beneficial for a high selectivity of H_2 (>40%)[1].

As plotted in Fig. 5.6, plasma-alumina reforming achieved the maximum hydrogen production rate and the lowest SER for hydrogen of 1.04 mmol/s and 13.7 kJ/mol, respectively. The addition of catalysts in plasma reformer noticeably promoted hydrogen productivity, which led to much lower required energy for

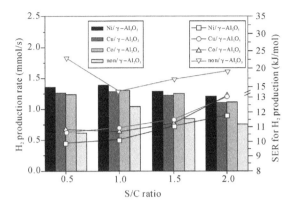

Fig. 5.6 Hydrogen production rates and SER for H_2 production with different catalytic beds in plasma-catalytic reforming

producing hydrogen of 1 mol, specially at a lower S/C ratio (0.5–1.0). For O/C = 0.5 and QEtOH = 0.10 g/s, the optimal H_2 productivity (2.06 mmol/s) and SER for H_2 (9.9 kJ/mol) were obtained at S/C = 0.5 with the Ni catalyst[1].

5.4 Conclusion

Though a good conversion rate and attractive SER for hydrogen production were obtained in this work, to further improve the performance, more attempts must be made to select a proper catalyst with a lower cost but better activity and that lasts a lifetime. Additionally, due to the high surface-area-to-volume ratio, miniaturization of the plasma reactor should be further investigated and modified to effectively generate and use the active species.

References

[1] Du CM, Ma DY, Wu J, Lin YC, Xiao W, Ruan JJ, Huang DW. Plasma-catalysis reforming for H_2 production from ethanol. Int J Hydrogen Energy. 2015; 40(45): 15381–15864.
[2] Rollier JD, Gonzalez-Aguilar J, Petitpas G, Darmon A, Fulcheri L, Metkemeijer R. Experimental study on gasoline reforming assisted by a nonthermal arc discharge. Energy Fuel. 2008; 22(1): 556–560.
[3] Gonzalez-Aguilar J, Petitpas G, Lebouvier A, Rollier JD, Darmon A, Fulcheri L. Three stages modeling of n-octane reforming assisted by a nonthermal arc discharge. Energy Fuel. 2009; 23: 4931–4936.
[4] Nozaki T, Okazaki K. Non-thermal plasma catalysis of methane: principles, energy efficiency, and applications. Catal Today. 2013; 211: 29–38.

Chapter 6
Mechanism for the Plasma Reforming of Ethanol

6.1 Mechanism Analysis of the Single Plasma Reforming of Ethanol

High energy electrons and active radicals are significant components of plasma chemistry. A non-thermal plasma arc reaction can be divided into two stages; the first stage is the electron bombardment reaction. In this stage, the electrons gain a lot of energy in the strong electric field and convert to high-energy electrons; later the electrons with higher energy bombard the molecules to make the molecular rupture of covalent bonding and generate small molecules and free radicals. The electrons with low energy can make molecules or atoms go into the excited state as well. The second stage is the radical reaction. Radicals and other particles (ground state or excited state of molecules, atoms, or other radicals) obtain the final product through a series of radical collisions. The following passage will analyze the mechanism for the single non-thermal arc plasma reforming of ethanol from three aspects: electron-molecule collision, radical reaction and generation and conversion of major products. It is worth mentioning that under the best reforming conditions of single non-thermal arc reforming, there is no important carbon generation phenomenon in the reactor wall and production stream and the NO_x content of the production stream is very low, so the inhibition of the generation of charcoal and NO_x caused by the non-thermal arc plasma reforming will also be discussed[1].

6.1.1 Electron-Molecule Collision

In the plasma reaction, the electrons are accelerated in the strong electric field and make elastic or inelastic collisions with other particles. In the elastic collisions, the electrons lose little energy, while in the inelastic collisions, the kinetic energy of the electrons is almost all transformed into the potential energy of heavy particles, so the heavy particles are excited, dissociated or ionized. Therefore, we can see that the average level of energy and density of the electrons make a significant impact through inelastic collisions.

The main reactions of the electron-molecule collision are as follows:

Excited reaction:
$$A + e \rightarrow A^* + e \tag{6.1}$$

Dissociation reaction:
$$AB + e \rightarrow A + B + e \tag{6.2}$$

Ionization reaction:
$$A + e \rightarrow A^+ + 2e \tag{6.3}$$

Dissociative ionization reaction:
$$AB + e \rightarrow A^+ + B + 2e \tag{6.4}$$

Complex reaction:
$$e + A^+ \rightarrow A \tag{6.5}$$

In the above reaction formula, A stands for heavy particle (it includes ground-state molecules or atoms), AB stands for molecule, e stands for electron, A^+ stands for positive ion and A* stands for the excited-state heavy particle. When e exists on both sides of the reaction formula, the left one is the high-energy electron, the right one loses almost all of its kinetic energy after the inelastic collision and changes into a low-energy electron. The low-energy electron can regain energy in the electric field and change into a high-energy one to keep the high-energy electrons at a higher density. Table 6.1 lists the main reactions of the electron-molecule collision in the plasma reforming of ethanol.

Table 6.1 The main reactions of the electron-molecule collision during the auto-thermal reforming of ethanol assisted by plasma

No.	Electron-molecule collision	Threshold energy (eV)	Rate constants (cm^3/s)	References
H_2O dissociation				
1	$H_2O + e \rightarrow H + OH + e$	7.00	3.6×10^{-10}	[2–3]
O_2 dissociation				
2	$O_2 + e \rightarrow O + O + e$	6.00	1.4×10^{-9}	[2, 4]
C_2H_5OH dissociation				
3	$C_2H_5OH + e \rightarrow CH_3CHOH + H + e$	7.82	1.0×10^{-9}	[2]

No.	Electron-molecule collision	Threshold energy (eV)	Rate constants (cm^3/s)	References
4	$C_2H_5OH+e \rightarrow C_2H_5+OH+e$	7.90	4.7×10^{-10}	[2]
5	$C_2H_5OH+e \rightarrow CH_3+CH_2OH+e$	7.38	1.8	[2]
CH_3CHOH dissociation				
6	$CH_3CHOH+e \rightarrow CH_3CHO+H+e$	8.80	1.0×10^{-9}	[2]
7	$CH_3CHOH+e \rightarrow OH+C_2H_4+e$	3.46	–	[5]
CH_3CHO dissociation				
8	$CH_3CHO+e \rightarrow CH_3CO+H+e$	7.60	3.9×10^{-9}	[2]
CH_3CO dissociation				
9	$CH_3CO+e \rightarrow CH_3+CO+e$	1.04	3.9×10^{-9}	[2]
CH_2OH dissociation				
10	$CH_2OH+e \rightarrow CH_2O+H+e$	3.18	–	[6]
CH_2O dissociation				
11	$CH_2O+e \rightarrow HCO+H+e$	7.56	4.1×10^{-9}	[2]
12	$CH_2O+e \rightarrow CO+H_2+e$	7.66	4.9×10^{-9}	[2]
HCO dissociation				
13	$HCO+e \rightarrow CO+H+e$	1.6	2.1×10^{-9}	[2]
C_2H_5 dissociation				
14	$C_2H_5+e \rightarrow C_2H_4+H+e$	3.38	1.6×10^{-10}	[2]
C_2H_4 *dissociation*				
15	$C_2H_4+e \rightarrow C_2H_3+H+e$	10.00	5.7×10^{-10}	[2, 6]
C_2H_3 dissociation				
16	$C_2H_3+e \rightarrow C_2H_2+H+e$	3.48	1.8×10^{-9}	[2]
CH_4 dissociation				
17	$CH_4+e \rightarrow CH_3+H+e$	4.50	2.8×10^{-9}	[2, 6]

In the plasma reforming of ethanol, ethanol, water and air that make up the reaction substrates all react like Eqs. (6.1)–(6.5) through collisions of high-energy electrons. Under the collisions of high-energy electrons, the water molecules are dissociated into H and OH radicals[7–8]:

$$H_2O + e \rightarrow H + OH + e \tag{6.6}$$

In addition, the strong electric field can dissociate water into H^+ and OH^- as well. The two kinds of ions above can produce H and OH radicals by absorbing or releasing electrons[7–8]:

$$H_2O \rightarrow H^+ + OH^- \tag{6.7}$$

$$OH^- \rightarrow OH + e \tag{6.8}$$

$$H^+ + e \rightarrow H \tag{6.9}$$

Among them, H radicals can take H atoms from some of the H-containing molecules and generate H_2, so improving its concentration is beneficial to improving the yield of H_2; OH radicals have a very strong activity as well and they can promote ethanol degradation through collision with ethanol. Thus the dissociation of the water molecules helps to improve the conversion of ethanol and hydrogen output. It should be noted that high-energy electrons can also be absorbed by strong negative electric water molecules easily and generate H_2O^-, thereby inhibiting the generation of O radical and adversely affecting the decomposition of ethanol[9].

$$H_2O + e \rightarrow H_2O^- \tag{6.10}$$

Air mainly includes O_2 and N_2, wherein the high-energy electrons bombard O_2 molecules and make the latter ones dissociated.

$$O_2 + e \rightarrow O + O + e \tag{6.11}$$

O radicals oxidize very strongly and their oxidizing is obviously higher than O_3. O radicals can react with the ethanol and radicals and generate an oxygen-containing compound. Or they can capture the H of the particles and generate OH radicals and can break the C=C bond of some C_2 products. High-energy electrons can also dissociate the N_2 molecules:

$$N_2 + e \rightarrow N + N + e \tag{6.12}$$

Since the N≡N bond energy is high in the N_2 (9.8 eV) while the bond energy of the O=O bond in O_2 and the H–O bond in H_2O is only 5.2 and 5.1 eV, compared to O_2 and H_2O, the dissociation of N_2 is more difficult and it does not participate much in the ethanol reforming.

Compared to H_2O, O_2, and N_2, the structure of the ethanol molecule is more complex, so it relates to the multi-stage decomposition process in the electron collision reactions. In ethanol reforming, the high-energy electron can be captured by the ethanol molecule and become a negative ion to break the C–C bonds, the C–O bond and the O–H bond of ethanol as a result of the increase in the internal repulsion[8]. The following electron-molecule reactions regarding ethanol will occur at first:

$$C_2H_5OH + e \rightarrow CH_3CHOH + H + e \qquad (6.13)$$

$$C_2H_5OH + e \rightarrow CH_3CHO + H_2 + e \qquad (6.14)$$

$$C_2H_5OH + e \rightarrow C_2H_5 + OH + e \qquad (6.15)$$

$$C_2H_5OH + e \rightarrow C_2H_4 + H_2O + e \qquad (6.16)$$

$$C_2H_5OH + e \rightarrow CH_3CH + H_2O + e \qquad (6.17)$$

$$C_2H_5OH + e \rightarrow CH_4 + CH_2O + e \qquad (6.18)$$

$$C_2H_5OH + e \rightarrow CH_3 + CH_2OH + e \qquad (6.19)$$

From the above reaction formulas we are able to know that in the first electron bombardment, the ethanol molecule produces radicals including CH_3CHOH, H, CH_3, CH_2OH, CH_3CHO, CH_2O, C_2H_5, OH and CH_3CH and small molecules including C_2H_4, H_2O, H_2 and CH_4. Later, high-energy electrons and the above particles have a further collision in the electric field and produce smaller particles or molecules. In the discharge area, the molecules generated by the degradation of the ethanol are related to the average energy level of the electrons. Fig. 6.1 shows the main electron-molecule reactions of the ethanol in a high-energy electron bombardment. In the fully-ionized stage and the highly-ionized stage, the average energy level of the electrons is high, so the ethanol bombarded by the electrons is degraded more thoroughly and the main product is the low molecular weight radical; in the weakly-ionized stage and the downstream of the discharge region, since the average energy level of the electrons is low, the level of the ethanol degradation is low and the product includes a greater number of high-molecular weight particles, and at this time, the low-molecular weight radicals generated in the fully-ionized and highly-ionized stages collide with each other and form stable molecules. As described in Chapter 5, the fully-ionized and highly-ionized stages take up most of the time in miniature non-thermal arc plasma, making sure of a high level of degradation of the ethanol. As shown in Fig. 6.1, a large

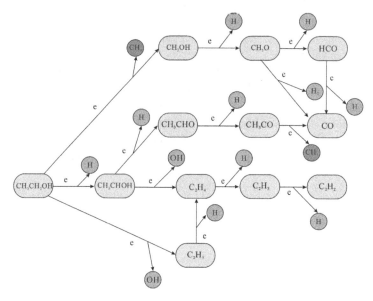

Fig. 6.1 The likely main dissociative pathways of ethanol induced by energetic electron bombardment

number of high-activity H, OH, CH_3 and OH radicals are generated in the collision of the electrons and the ethanol. CH_3COOH has three successive pathways of degradation under the bombardment of the electrons.

In the first pathway, the C=C bond breaks first and the ethanol produces CH_3 and CH_2OH radicals and then the CH_2OH radicals remove the hydrogen atoms to generate formaldehyde (CH_2O). Since the stability of the ethanol is better than that of the formaldehyde in the non-thermal discharge environment, the formaldehyde continues to dehydrogenate and generates HCO and finally stabilizes in the form of CO[10]:

$$C_2H_5OH + e \rightarrow CH_3 + CH_2OH + e \quad (6.19)$$

$$CH_2OH + e \rightarrow CH_2O + H + e \quad (6.20)$$

$$CH_2O + e \rightarrow HCO + H + e \quad (6.21)$$

$$HCO + e \rightarrow CO + H + e \quad (6.22)$$

$$CH_2O + e \rightarrow CO + H_2 + e \quad (6.23)$$

In the second pathway, the C=C bond does not break at first, and the ethanol removes an H radical to generate CH_3CHOH and later it removes another H radical to generate acetaldehyde. If the acetaldehyde continues to break down under the bombardment of the electrons, then it can remove an H and a CH_3 radical one after

6 Mechanism for the Plasma Reforming of Ethanol

another and obtain a stable molecule CO. This pathway is essentially one of the main sources of liquid product acetaldehyde (bombarded by radicals such as H, OH, O, the ethanol can undergo two dehydrogenation reactions consecutively and produce acetaldehyde). The H_2 content in the gas obtained from the above process can be up to a high level (such as higher than 80%) while the conversion, utilization and energy efficiency of the ethanol show weakness. Under normal pressure and low temperature, the main products CH_3CHO and H_2 through non-thermal glow discharge reforming are obtained[11]:

$$C_2H_5OH + e \rightarrow CH_3CHOH + H + e \quad (6.13)$$

$$CH_3CHOH + e \rightarrow CH_3CHO + H + e \quad (6.24)$$

$$CH_3CHO + e \rightarrow CH_3CO + H + e \quad (6.25)$$

$$CH_3CO + e \rightarrow CH_3 + CO + e \quad (6.26)$$

The above two pathways are both related to the generation of the free radical CH_3, which is the main source of the products CH_4 and C_2H_6:

$$CH_3 + H \rightarrow CH_4 \quad (6.27)$$

$$CH_3 + CH_3 \rightarrow C_2H_6 \quad (6.28)$$

In the third pathway, the ethanol either removes the H radical first and then the OH radical or removes the OH radical first and then the H radical; the two ways both generate C_2H_4. In the process of removing the OH radical first, the C_2H_5 radical is produced, which has some significance for the byproducts C_2H_2 and C_2H_4. In addition, in this process the C=C bond does not break, so it is one of the main ways to produce the by-products of C_2 hydrocarbon:

$$C_2H_5OH + e \rightarrow CH_3CHOH + H + e \quad (6.13)$$

$$CH_3CHOH + e \rightarrow OH + C_2H_4 + e \quad (6.18)$$

$$C_2H_5OH + e \rightarrow C_2H_5 + OH + e \quad (6.15)$$

$$C_2H_5 + e \rightarrow C_2H_4 + H + e \quad (6.29)$$

$$C_2H_4 + e \rightarrow C_2H_3 + H + e \quad (6.30)$$

$$C_2H_3 + e \rightarrow C_2H_2 + H + e \quad (6.31)$$

Ethanol generates a large number of H, OH and CH_3 radicals in the electronmolecule reaction. H_2O molecules and O_2 molecules also generate OH, H radicals and O radicals in the electron impact dissociation. In addition, as the main product, CH_4 can generate methyl radicals (CH_3) under the hit of high-energy electrons:

$$CH_4 + e \rightarrow CH_3 + H + e \tag{6.32}$$

In summary, the main types of radicals generated in non-thermal arc reforming are H, OH, O and CH_3 radicals. The high-energy electron excitation effect on the reforming substrate can also be effective in promoting the ethanol degradation and product formation. According to the Arrhenius equation, the Chemical reaction rate is expressed as:

$$k = A \cdot \exp\left(-\frac{E_a}{RT}\right) \tag{6.33}$$

In the equation, k is the chemical reaction rate, A is the pre-exponential factor, T is the absolute temperature, R is the molar gas constant, and E_a is the activation energy. When the ground-state particle jumps into the excited state in the collisions of the electron-molecule, the related activation energy decreases with the improvement of the energy level of the particles participating in the reaction. Assuming that the average energy increment of ground-state particles in the excitation reaction is E_p, then the activation energy is reduced to ($E_a - E_p$), while in the non-equilibrium plasma $E_p \gg RT$ ($RT < 0.1$ eV), in other words, $E_p/RT \gg 1$. In this case, the reaction rate is as follows[12]:

$$\begin{aligned} K_p &= A \cdot \exp\left(-\frac{E_a - E_p}{RT}\right) \\ &= A \cdot \exp\left(-\frac{E_a}{RT}\right) \cdot \exp\left(\frac{E_p}{RT}\right) \\ &= K \exp\left(\frac{E_p}{RT}\right) \end{aligned} \tag{6.34}$$

In the above equation, k and k_p respectively stand for the reaction rate of the ground-state and the excited state particles, which shows that the reaction rate can be increased to the (E_p/RT) times of the original.

6.1.2 Free Radical Reaction

High-energy electrons and molecules can produce excited-state particles, free radicals and charged ions when inelastic collisions occur. However, the life of the excited-state particles is from sub-nanosecond to microsecond, so its role in promoting the reaction is limited in the gas phase reaction conditions[13]. The electron bombardment needs an electron energy level which is higher than the molecular dissociation energy to successfully cause ionization reactions. Therefore, in addition to high-energy electrons, the major participants in the plasma chemical reaction are free radicals whose life is from a sub-nanosecond to a microsecond[14–16]. Besides, the structure of the convergent-divergent metal tube used in the research is the Laval nozzle, which can improve the airflow velocity to the level of subsonic speed and even supersonic, making the air pressure at the back of the convergent-divergent tube decrease abruptly in order to extend the duration of the activity of the free radicals[17]. As described above, a large number of H, OH, O and CH_3 radicals produced in the process of electron-molecule reaction and the interaction of these radicals and other particles have very important effects on the generation of the reformate. Table 6.2 shows the main reactions and the products of the four above radicals in the process of the non-thermal arc discharge reforming of ethanol, while Table 6.3 also provides the reactions of each radical and the main reactions pyrolysis of molecules and molecular fragments caused by the third body collision. The H radical is mainly obtained through high-energy electrons bombarding water molecules and ethanol (see Eqs. (6.6) and (6.13)) and high-energy electrons can generate H radicals by continuing to bombard the molecular fragments of the ethanol. This radical is of decisive significance for hydrogen production. Two H radicals can generate hydrogen molecules through complex reactions:

$$H + H \xrightarrow{M, H_2, H_2O} H_2 \qquad (6.35)$$

Table 6.2 The main radical reactions and products during ethanol reforming assisted by non-thermal plasma[49]

	H radical	O radical	OH radical	CH_3 radical
H radical	H_2	OH	H_2O	$^aCH_4$ $^bCH_2 + H_2$
O radical	OH	O_2	$O_2 + H$	$CH_2O + H$
OH radical	H_2O	$O_2 + H$	$CH_2O + H$	$CH_2 + H_2O$

	H radical	O radical	OH radical	CH$_3$ radical
CH$_3$ radical	^{a}CH$_4$ ^{b}CH$_2$ + H$_2$	CH$_2$O + H	^{a}CH$_3$OH ^{b}CH$_2$ + H$_2$O	C$_2$H$_6$
CH$_3$CH$_2$OH	CH$_3$CHOH + H$_2$	CH$_3$CHOH + OH	CH$_3$CHOH + H$_2$O	CH$_4$ + CH$_3$CHOH
CH$_2$OH	^{a}CH$_2$O + H$_2$ ^{b}CH$_3$ + OH	CH$_2$O + OH	CH$_2$O + H$_2$O	–
CH$_2$O	^{a}HCO + H$_2$ ^{b}CH$_3$O	HCO + OH	HCO + H$_2$O	–
HCO	CO + H$_2$	^{a}CO + OH ^{a}CO$_2$ + H	CO + H$_2$O	–
CH$_3$CHOH	^{a}C$_2$H$_4$ + H$_2$O ^{a}CH$_3$ + CH$_2$OH	CH$_3$CHO + OH	CH$_3$CHO + H$_2$O	–
CH$_3$CHO	CH$_3$CO + H$_2$	CH$_3$CO + OH	^{a}HCOOH + CH$_3$ ^{b}CH$_3$CO + H$_2$O	CH$_3$CO + CH$_4$
C$_2$H$_6$	C$_2$H$_5$ + H$_2$	C$_2$H$_5$ + OH	C$_2$H$_5$ + H$_2$O	C$_2$H$_5$ + CH$_4$
C$_2$H$_5$	^{a}CH$_3$ + CH$_3$ ^{a}C$_2$H$_6$ ^{b}C$_2$H$_4$ + H$_2$	CH$_3$ + CH$_2$O	C$_2$H$_4$ + H$_2$O	–
C$_2$H$_4$	^{a}C$_2$H$_5$ ^{b}C$_2$H$_3$ + H$_2$	^{a}CH$_3$ + HCO ^{b}C$_2$H$_3$ + H$_2$O	C$_2$H$_3$ + H$_2$O	C$_2$H$_3$ + CH$_4$
C$_2$H$_3$	C$_2$H$_4$	–	C$_2$H$_2$ + H$_2$O	–
C$_2$H$_2$	–	CH$_2$ + CO	CH$_3$ + CO	–
CH$_4$	CH$_3$ + H$_2$	CH$_3$ + OH	CH$_3$ + H$_2$O	–
H$_2$O	OH + H$_2$	OH + OH	–	–
CO	–	CO$_2$	CO$_2$ + H	–

^{a} and ^{b} refer to the major product and the minor product respectively, which can be distinguished by the theoretical values of k (reaction rate coefficient) of the reaction temperature in the range of 300–2300 K and the expression of k is derived from [18–19]

6 Mechanism for the Plasma Reforming of Ethanol

Table 6.3 The main radical reactions and dissociative reactions induced by the third body in non-thermal plasma discharge

No.	Radical reaction formula	reaction rate coefficient (cm³/s)	References
Reactions related to H radical			
1	$H + H + M \rightarrow H_2 + M$	$2.76 \times 10^{-30} \times T^{-1.00}$	[19–20]
2	$H + H + H_2 \rightarrow H_2 + H_2$	$2.54 \times 10^{-31} \times T^{-0.60}$	[19–20]
3	$H + H + H_2O \rightarrow H_2 + H_2O$	$1.65 \times 10^{-28} \times T^{-1.25}$	[19–20]
4	$H + O + M \rightarrow OH + M$	$1.30 \times 10^{-29} \times T^{-1.00}$	[19–20]
5	$H + OH + M \rightarrow H_2O + M$	$6.09 \times 10^{-26} \times T^{-2.00}$	[19–20]
6	$CH_3 + H(+M) \rightarrow CH_4(+M)$	$3.55 \times 10^{-9} \times T^{-0.40}$	[19, 21]
7	$CH_3 + H \rightarrow CH_2 + H_2$	$1.49 \times 10^{-10} \times \exp\left(\frac{-7602}{T}\right)$	[19, 22]
8	$C_2H_5OH + H \rightarrow CH_3CHOH + H_2$	$4.28 \times 10^{-17} \times T^{1.65} \times \exp\left(\frac{-1423}{T}\right)$	[19, 21]
9	$CH_2OH + H \rightarrow CH_2O + H_2$	3.32×10^{-11}	[19, 22]
10	$CH_2OH + H \rightarrow CH_3 + OH$	1.66×10^{-11}	[19]
11	$CH_2O + H(+M) \rightarrow HCO + H_2(+M)$	$8.97 \times 10^{-13} \times T^{0.45} \times \exp\left(\frac{-1812}{T}\right)$	[19, 23]
12	$CH_2O + H(+M) \rightarrow CH_3O(+M)$	$8.97 \times 10^{-13} \times T^{0.45} \times \exp\left(\frac{-1309}{T}\right)$	[19, 23]
13	$HCO + H \rightarrow CO + H_2$	$1.98 \times 10^{-11} \times T^{0.25}$	[19, 22]
14	$CH_3CHOH + H \rightarrow C_2H_4 + H_2O$	4.98×10^{-11}	[19, 21]
15	$CH_3CHOH + H \rightarrow CH_3 + CH_2OH$	4.98×10^{-11}	[19, 21]
16	$CH_3CHO + H \rightarrow CH_3CO + H_2$	$7.74 \times 10^{-11} \times T^{-0.35} \times \exp\left(\frac{1504}{T}\right)$	[19, 21]
17	$C_2H_6 + H \rightarrow C_2H_5 + H_2$	$1.91 \times 10^{-16} \times T^{1.90} \times \exp\left(\frac{-3791}{T}\right)$	[19, 23]
18	$C_2H_5 + H \rightarrow 2CH_3$	4.98×10^{-11}	[19, 22]
19	$C_2H_5 + H \rightarrow C_2H_6$	4.98×10^{-11}	[19, 21]
20	$C_2H_5 + H \rightarrow C_2H_4 + H_2$	$2.08 \times 10^{-10} \times \exp\left(\frac{-4027}{T}\right)$	[19–20]
21	$C_2H_4 + H(+M) \rightarrow C_2H_5(+M)$	$8.97 \times 10^{-13} \times T^{0.45} \times \exp\left(\frac{-916}{T}\right)$	[19, 23]

(To be continued)

(Table 6.3)

No.	Radical reaction formula	reaction rate coefficient (cm^3/s)	References
22	$C_2H_4 + H \rightarrow C_2H_3 + H_2$	$2.20 \times 10^{-18} \times T^{2.53} \times \exp\left(\frac{-6162}{T}\right)$	[19, 23]
23	$C_2H_3 + H(+M) \rightarrow C_2H_4(+M)$	$1.01 \times 10^{-11} \times T^{0.27} \times \exp\left(\frac{-141}{T}\right)$	[19, 23]
24	$C_2H_3 + H \rightarrow C_2H_2 + H_2$	1.49×10^{-10}	[19, 22]
25	$CH_4 + H \rightarrow CH_3 + H_2$	$3.65 \times 10^{-20} \times T^{3.00} \times \exp\left(\frac{-4405}{T}\right)$	[19, 21]
26	$H_2O + H \rightarrow OH + H_2$	$7.5 \times 10^{-16} \times T^{1.60} \times \exp\left(\frac{-9030}{T}\right)$	[18]
Reactions related to O radical			
4	$H + O + M \rightarrow OH + M$	$1.30 \times 10^{-29} \times T^{-1.00}$	[19–20]
27	$O + O + M \rightarrow O_2 + M$	$5.21 \times 10^{-35} \times \exp\left(\frac{900}{T}\right)$	[19–20]
28	$O + OH \rightarrow O_2 + H$	$2.0 \times 10^{-10} \times T^{-0.352} \times \exp\left(\frac{113}{T}\right)$	[18]
29	$O + CH_3 \rightarrow CH_2O + H$	1.33×10^{-10}	[19–20]
30	$CH_3CH_2OH + O \rightarrow CH_3CHOH + OH$	$3.12 \times 10^{-17} \times T^{1.85} \times \exp\left(\frac{-918}{T}\right)$	[19, 21]
31	$CH_2OH + O \rightarrow CH_2O + OH$	1.66×10^{-11}	[19, 22]
32	$CH_2O + O \rightarrow HCO + OH$	$2.99 \times 10^{-11} \times \exp\left(\frac{-1551}{T}\right)$	[19, 22]
33	$HCO + O \rightarrow CO + OH$	4.98×10^{-11}	[19, 22]
34	$HCO + O \rightarrow CO_2 + H$	4.98×10^{-11}	[19, 22]
35	$CH_3CHOH + O \rightarrow CH_3CHO + OH$	1.66×10^{-11}	[19–20]
36	$CH_3CHO + O \rightarrow CH_3CO + OH$	$2.94 \times 10^{-6} \times T^{-1.90} \times \exp\left(\frac{1498}{T}\right)$	[19–20]
37	$C_2H_6 + O \rightarrow C_2H_5 + OH$	$4.98 \times 10^{-17} \times T^{2.00} \times \exp\left(\frac{-2575}{T}\right)$	[19, 22]
38	$C_2H_5 + O \rightarrow CH_3 + CH_2O$	1.66×10^{-10}	[19, 22]
39	$C_2H_4 + O \rightarrow CH_3 + HCO$	$1.69 \times 10^{-17} \times T^{1.88} \times \exp\left(\frac{-90}{T}\right)$	[19, 22]
40	$C_2H_4 + O \rightarrow C_2H_3 + H_2O$	$5.63 \times 10^{-18} \times T^{1.88} \times \exp\left(\frac{-90}{T}\right)$	[19, 22]
41	$C_2H_2 + O \rightarrow CH_2 + CO$	$1.02 \times 10^{-17} \times T^{2.00} \times \exp\left(\frac{-957}{T}\right)$	[19–20]

(To be continued)

(Table 6.3)

No.	Radical reaction formula	reaction rate coefficient (cm³/s)	References
42	$CH_4 + O \rightarrow CH_3 + OH$	$1.15 \times 10^{-15} \times T^{1.56} \times \exp\left(\frac{-4272}{T}\right)$	[19–20]
43	$CO + O + M \rightarrow CO_2 + M$	$1.70 \times 10^{-33} \times \exp\left(\frac{-1510}{T}\right)$	[19, 22]
Reactions related to OH radical			
5	$H + OH + M \rightarrow H_2O + M$	$6.09 \times 10^{-26} \times T^{-2.00}$	[19–20]
28	$O + OH \rightarrow O_2 + H$	$2.0 \times 10^{-10} \times T^{-0.352} \times \exp\left(\frac{113}{T}\right)$	[18]
44	$OH + OH \rightarrow H_2O + O$	$5.93 \times 10^{-20} \times T^{2.40} \times \exp\left(\frac{1063}{T}\right)$	[19–20]
45	$CH_3 + OH(+M) \rightarrow CH_3OH(+M)$	$1.44 \times 10^{-10} \times T^{0.10}$	[19–22]
46	$CH_3CH_2OH + OH \rightarrow CH_3CHOH + H_2O$	$7.70 \times 10^{-13} \times T^{0.15}$	[19, 21]
47	$CH_2OH + OH \rightarrow CH_2O + H_2O$	1.66×10^{-11}	[19, 22]
48	$CH_2O + OH \rightarrow HCO + H_2O$	$5.70 \times 10^{-15} \times T^{1.18} \times \exp\left(\frac{225}{T}\right)$	[19, 24]
49	$HCO + OH \rightarrow CO + H_2O$	1.66×10^{-10}	[19, 22]
50	$CH_3CHOH + OH \rightarrow CH_3CHO + H_2O$	8.30×10^{-12}	[19, 21]
51	$CH_3CHO + OH \rightarrow HCOOH + CH_3$	$1.53 \times 10^{-17} \times T^{1.50} \times \exp\left(\frac{484}{T}\right)$	[19, 21]
52	$CH_3CHO + OH \rightarrow CH_3CO + H_2O$	$4.98 \times 10^{-9} \times T^{-1.08}$	[19, 22]
53	$C_2H_6 + OH \rightarrow C_2H_5 + H_2O$	$1.20 \times 10^{-17} \times T^{2.00} \times \exp\left(\frac{-435}{T}\right)$	[19, 22]
54	$C_2H_5 + OH \rightarrow C_2H_4 + H_2O$	6.64×10^{-11}	[19–20]
55	$C_2H_4 + OH \rightarrow C_2H_3 + H_2O$	$3.35 \times 10^{-11} \times \exp\left(\frac{-2988}{T}\right)$	[19, 25]
56	$C_2H_3 + OH \rightarrow C_2H_2 + H_2O$	3.32×10^{-11}	[19, 22]
57	$C_2H_2 + OH \rightarrow CH_3 + CO$	$8.02 \times 10^{-28} \times T^{4.00} \times \exp\left(\frac{1007}{T}\right)$	[19, 22]
58	$CH_4 + OH \rightarrow CH_3 + H_2O$	$6.96 \times 10^{-18} \times T^{2.00} \times \exp\left(\frac{-1282}{T}\right)$	[19–20]
59	$CO + OH \rightarrow CO_2 + H$	$1.56 \times 10^{-20} \times T^{2.25} \times \exp\left(\frac{1184}{T}\right)$	[19, 21]

(To be continued)

(Table 6.3)

No.	Radical reaction formula	reaction rate coefficient (cm³/s)	References
Reactions related to OH_3 radical			
6	$CH_3 + H(+M) \rightarrow CH_4(+M)$	$3.55 \times 10^{-9} \times T^{-0.40}$	[19, 21]
29	$O + CH_3 \rightarrow CH_2O + H$	1.33×10^{-10}	[19–20]
45	$CH_3 + OH(+M) \rightarrow CH_3OH(+M)$	$1.44 \times 10^{-10} \times T^{0.10}$	[19, 22]
60	$CH_3 + CH_3(+M) \rightarrow C_2H_6(+M)$	$1.53 \times 10^{-7} \times T^{-1.17} \times \exp\left(\frac{320}{T}\right)$	[19, 21]
61	$CH_3CH_2OH + CH_3 \rightarrow CH_3CHOH + CH_4$	$1.21 \times 10^{-21} \times T^{2.99} \times \exp\left(\frac{-4001}{T}\right)$	[19, 21]
62	$CH_3CHO + CH_3 \rightarrow CH_3CO + CH_4$	$6.48 \times 10^{-31} \times T^{5.80} \times \exp\left(\frac{-1108}{T}\right)$	[19, 21]
63	$C_2H_6 + CH_3 \rightarrow C_2H_5 + CH_4$	$9.13 \times 10^{-25} \times T^{4.00} \times \exp\left(\frac{-4179}{T}\right)$	[19, 22]
64	$C_2H_4 + CH_3 \rightarrow C_2H_3 + CH_4$	$1.10 \times 10^{-23} \times T^{3.70} \times \exp\left(\frac{-4783}{T}\right)$	[19–20]
Decomposition reactions of collision with third body			
66	$CH_3CH_2OH(+M) \rightarrow CH_5 + OH(+M)$	$1.25 \times 10^{23} \times T^{-1.54} \times \exp\left(\frac{48332}{T}\right)$	[19, 21]
67	$CH_3CH_2OH(+M) \rightarrow C_2H_4 + H_2O(+M)$	$2.79 \times 10^{13} \times T^{0.09} \times \exp\left(\frac{-33295}{T}\right)$	[19, 21]
68	$CH_3CH_2OH(+M) \rightarrow CH_3CHO + H_2(+M)$	$7.24 \times 10^{11} \times T^{0.10} \times \exp\left(\frac{-45816}{T}\right)$	[19, 21]
69	$CH_3CHOH + M \rightarrow CH_3CHO + H + M$	$1.66 \times 10^{-10} \times \exp\left(\frac{-12586}{T}\right)$	[19, 21]
70	$C_2H_4(+M) \rightarrow CH_2 + H_2(+M)$	$1.80 \times 10^{14} \times \exp\left(\frac{-43799}{T}\right)$	[19, 22]

H radicals can capture the hydrogen of ethanol to obtain the hydrogen molecule. This is the main source of early hydrogen production through the degradation of ethanol[5].

$$C_2H_5OH + H \rightarrow CH_3CHOH + H_2 \quad (6.36)$$

Most of the hydrogen production reactions of the latter reactions are related to the hydrogen abstraction reaction of H radicals. As shown in Table 6.2, the H radical can continue to react with the molecular fragments of ethanol CH_2OH, CH_2O, HCO, CH_2HCO, C_2H_6, CH_4 and H_2O particles to generate hydrogen. It should be noted that the OH radical can react with H radicals and generate H_2O and the CH_3 radical can react with the H radical and generate CH_4 as well, so CH_3 and OH radicals have negative impacts on the hydrogen selectivity although they can help to improve the degradation rate of the ethanol.

The O radical is obtained mainly through high-energy electron bombarding O_2 molecules (see Eq. (6.11)). This free radical has a strong oxidation and is typically able to capture a hydrogen atom from molecules or radicals and generate OH radicals as well as breaking the C=C of C_2 particles such as C_2H_4 and C_2H_5 to get CH_3 radicals. In addition, the O radical and the O_2 molecule can generate O_3 through the three-body collision reaction. Although O_3 has a high oxidation resistance and stability as well, its activity is far weaker than the O radical.

$$O + O_2 + M \rightarrow O_3 + M \quad (6.37)$$

The OH radical is obtained mainly through high-energy electron bombarding water, ethanol and molecular fragments of ethanol (see Eqs. 6.6, 6.13 and Fig. 6.1). Besides, the O radical and molecules or molecular fragments can also generate OH radicals through dehydrogenation. The CH_3 radical is generated through the breaking of the C=C bond. The OH, CH_3 radical can also have dehydrogenation with other particles generally and obtain H_2O and CH_4 respectively, so the CH_3 radical is the main precursor of the product CH_4.

In summary, the amount and intensity of the H radical have a decisive influence on the yield and selectivity of H_2, while the O radical, the OH radical and the CH_3 radical have a dehydrogenating effect, which will help to improve the conversion level of the ethanol but will not help to improve the H_2 selectivity. Furthermore, because of the strong oxidizing, O can break the C=C bond and obtain CH_3 radicals.

6.1.3 The Generation and Conversion of the Main Products

The main products of non-thermal arc reforming include H_2, CH_4, CO and CO_2, while the minor products include C_2H_2, C_2H_4, C_2H_6 and CH_3CHO, etc. As is known, the selectivity of the particular product is associated with its yield, which generally reflects the balance of the generation and conversion of the products. The improvement of the selectivity of the specified product can be achieved by promoting the conversion of other particles into the product or preventing the conversion of the product into other products. Discussion of the generation and conversion of each product helps to fundamentally improve the selectivity of the specified product and achieve the best reforming performance.

The main products containing hydrogen in the experiment are H_2 and CH_4, so hydrogen-producing reactions and methane-producing reactions compete for the hydrogen atom during the reforming. That is to say, with the same ethanol degradation rate, in order to increase the hydrogen yield, the inhibition of reactions producing methane or the promotion of the degradation of CH_4 is needed. The ways of producing hydrogen and methane in the reforming process are shown in Fig. 6.2. As shown in Fig. 6.2, the H radical and the CH_3 radical play a role in the control of the production of H_2 and CH_4, respectively. As mentioned above, the selectivity of C_2H_2, C_2H_4 and C_2H_6 is low in the reforming process, and the main route for the production of hydrogen is shown in the dashed box. During the process of ethanol degradation, as long as the C=C bond breaks, CH_3 will be unavoidably generated, and CH_4 will also be produced by the collisions of the CH_3 radicals with some molecules and molecular fragments including H radicals. In the reaction of

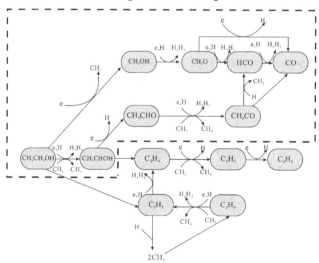

Fig. 6.2 Competition between hydrogen-generating reactions and methane-generating reactions within non-thermal plasma reforming processes

6 Mechanism for the Plasma Reforming of Ethanol

CH$_3$CHO converting to CH$_3$CO, competition between hydrogen-producing reactions and methane-producing reactions occurs and in the process of CH$_3$CO transforming into HCOCH$_3$CO, H radicals are consumed and CH$_3$ radicals are produced. Based on the above reasons, large amounts of CH$_4$ will be produced in the reforming process:

CH$_3$ radicals consume H directly and generate CH$_4$:

$$CH_3 + H \rightarrow CH_4 \tag{6.38}$$

Competition between H$_2$ and CH$_4$ occurs during the decomposition processes of CH$_3$CHO:

$$CH_3CHO + e \rightarrow CH_3CO + H + e \tag{6.25}$$

$$CH_3CHO + H \rightarrow CH_3CO + H_2 \tag{6.39}$$

$$CH_3CHO + CH_3 \rightarrow CH_3CO + CH_4 \tag{6.40}$$

CH$_3$CO consumes H and generates CH$_3$ radical:

$$CH_3CO + H \rightarrow CH_3 + HCO \tag{6.41}$$

Therefore, it is difficult to improve the selectivity of hydrogen by inhibiting the generation of the CH$_3$ radical and CH$_4$. It is necessary to regulate the conditions of the reaction and coordinate all potential reactions by thermodynamics. A more practical solution is to force CH$_4$ to further degrade and release hydrogen. Currently, various plasma technologies and catalyst processes have been applied for the methane reforming for the production of hydrogen. Similar to ethanol reforming, CH$_4$ reforming includes partial oxidation reforming, steam reforming and CO$_2$ reforming; the related reaction formulas are as follows:

Partial oxidation of methane:

$$2CH_4 + \frac{1}{2}O_2 \rightarrow 2CO + 2H_2O \quad \Delta H = -36 \text{ kJ/mol} \tag{6.42}$$

Steam reforming of methane:

$$CH_4 + H_2O \rightarrow CO_2 + 3H_2 \quad \Delta H = 206 \text{ kJ/mol} \tag{6.43}$$

Dry reforming of methane:

$$CH_4 + CO_2 \rightarrow 2CO + 2H_2 \quad \Delta H = 247 \text{ kJ/mol} \tag{6.44}$$

This shows that an appropriate amount of oxygen and water methane degradation have a certain role of promoting methane degradation. However, according to the discussions in Chaps. 3 and 4, we should also face up to the fact that adding too much oxygen and water will bring negative effects on the reforming, adding too much oxygen may lead to the oxidation of H_2 to H_2O; and adding too much water and air will cause the loss of more energy, lowering the degradation rate of the reaction substrate and the energy efficiency of the reforming.

There are large amounts of CO and CO_2 in the gas generated in the reforming. According to Fig. 6.1, Tables 6.2 and 6.3, CO comes mainly from the decomposition of HCO and CH_3CO, and acetylene can produce CO when oxidized by OH or O radicals:

Degradation of HCO:

$$HCO + e \rightarrow H + CO + e \tag{6.45}$$

$$HCO + H \rightarrow H_2 + CO \tag{6.46}$$

$$HCO + O \rightarrow OH + CO \tag{6.47}$$

$$HCO + OH \rightarrow H_2O + CO \tag{6.48}$$

Degradation of CH_3CO:

$$CH_3CO + e \rightarrow CH_3 + CO + e \tag{6.49}$$

Oxidation degradation of C_2H_2:

$$C_2H_2 + OH \rightarrow CH_3 + CO \tag{6.50}$$

$$C_2H_2 + O \rightarrow CH_2 + CO \tag{6.51}$$

Wherein C_2H_4 can be generated by CH_2 from the degradation of C_2H_2 through a complex reaction.

$$2CH_2 \rightarrow C_2H_4 \tag{6.52}$$

And CO can also be converted to CO_2 through an oxidation reaction and the water gas shift reaction:

Oxidation of CO:

$$CO + O \rightarrow CO_2 \tag{6.53}$$

$$CO + \frac{1}{2}O_2 \rightarrow CO_2 \tag{6.54}$$

Water gas shift:

$$CO + H_2O \rightarrow CO_2 + H_2 \qquad (6.55)$$

The water gas shift is a very important reaction in the reforming of ethanol, which can eliminate the CO that is the toxic platinum electrode of the fuel cell and improve the yield of hydrogen. This reaction usually takes place at a low temperature (<673 K); the temperature of the tail gas flow measured in this experiment was 120–210 °C, so it was helpful in promoting the production of hydrogen. Formulas (6.42) and (6.54) show that CO and CO_2 are products of CH_4 in different stages of oxidation, so the output ratio of CO, CO_2 and CH_4 can evaluate the overall degree of oxidation in the reforming process. Improvement of the oxidation level is usually accompanied by the increase of the ethanol conversion rate. A certain degree of oxidation helps to increase the output of H_2; however, excessive oxidation means that some H_2 will convert into H_2O, which will offset the output bonus of H_2 that the higher conversion rate of ethanol brings.

6.1.4 Suppression and Removal of Carbon Deposition in the Reforming Process

Carbon deposition is regarded as an important index to evaluate the effects of reforming in the traditional catalytic reforming of ethanol. There are two main degradation pathways of ethanol; one is to remove the hydrogen molecules and generate CH_3CHO, while the other one is to remove water molecules and generate C_2H_4[26].

Dehydrogenation reaction of ethanol:

$$CH_3CH_2OH \leftrightarrow CH_3CHO + H_2 \qquad (6.56)$$

Dehydration reaction of ethanol:

$$CH_3CH_2OH \leftrightarrow C_2H_4 + H_2O \qquad (6.57)$$

Among these reactions, C_2 products represented by C_2H_4 often form polymers on the wall, and thus result in the formation of a carbon deposition:

$$C_2H_4(a) - polymer - carbon\ deposition \qquad (6.58)$$

In addition, ways of generating carbon deposition also include the Boudouard reaction, the carbon gasification reverse reaction and the degradation of hydrocarbons.

The lower reaction temperature is in favor of the Boudouard reaction and the reverse reaction of carbon gasification while the higher temperature helps to improve the degradation of hydrocarbons to generate carbon deposition[26]:

The Boudouard reaction:

$$2CO \leftrightarrow CO_2 + C \qquad (6.59)$$

The carbon reverse reaction of gasification:

$$CO + H_2 \leftrightarrow H_2O + C \qquad (6.60)$$

The degradation of hydrocarbons:

$$CH_4 \leftrightarrow 2H_2 + C \qquad (6.61)$$

$$C_2H_4 \leftrightarrow 2C + 2H_2 \qquad (6.62)$$

Carbon deposition also affects the performance of the plasma reactor. Since carbon has an excellent electrical conductivity, the carbon attached to the electrodes will change the roughness of the surface of the electrodes, thereby changing the electrode gap, the discharge voltage and other operating conditions, and affect the stability of the reactor[27]. Moreover, when adding a catalyst layer in the discharge region or in the downstream of the discharge region, the solid carbon generated in the discharge region will attach to the surface or pore of the catalyst with the stream of materials, thus causing the deactivation of the catalyst. Wang et al. in Taiwan, China, have tried to add Ni/Al_2O_3 particles to the catalyst layer in the thermal plasma discharge region. After that, it was found that methane conversion and hydrogen selectivity improved to some extent. However, the improvement process lasted only 3–5 min, because a large number of nanocarbon particles deposited on the surface of the catalyst and the pore[28]. To sum up, anti-coking capacity is very significant in the plasma reaction process.

Since the temperature in the environment of non-thermal plasma is low, ways of producing carbon deposition that may exist are the dehydration of ethanol to generate C_2H_4, the Boudouard reaction and the degradation of hydrocarbons, but do not include the degradation of hydrocarbons. However, in practice, there is no notable carbon deposition on the wall of the reactor. Presumably, mechanisms that eliminate carbon deposition in miniature non-thermal arc reforming are the following:

1. Production of carbon deposition

We can inhibit the dehydration reaction of the ethanol molecules by selecting a suitable catalyst or carrier under certain catalytic reforming conditions. Since high-energy electron-molecule collisions have non-selectivity, the ethanol molecules will produce C_2H_4 unavoidably in the electron bombardment and radical reactions. As mentioned above, O_2 and H_2O of the reforming substrates produce a large number of O and H radicals (see Formulas 6.6 and 6.11) under the bombardment of high-energy electrons. Besides, the ethanol molecules also produce a certain amount of H radicals. O radicals have a very high oxidation potential, which can directly break the C=C bond of C_2H_4:

$$C_2H_4 + O \leftrightarrow CH_3 + HCO \tag{6.63}$$

Then CH_3 continues to be transformed into CH_4 while HCOO is transformed into H_2, CO and CO_2 via reactions (6.45–6.48).

In addition, H radicals also contribute to the transformation of C_2H_4 to C_1 product:

$$C_2H_4 + H \leftrightarrow C_2H_5 \tag{6.64}$$

$$C_2H_5 + H \leftrightarrow 2CH_3 \tag{6.65}$$

$$C_2H_5 + O \leftrightarrow CH_3 + CH_2O \tag{6.66}$$

2. The elimination of carbon deposition

O_2 and H_2O molecules can generate O and OH radicals, respectively, under the bombardment of high-energy electrons. Both the above radicals are of importance for the elimination of carbon[29]:

$$C + O \leftrightarrow CO \tag{6.67}$$

$$C + OH \rightarrow CO + H \tag{6.68}$$

Besides, steam can also generate hydrogen by reacting with free carbon atoms.

$$C + H_2O \rightarrow CO + H_2 \tag{6.69}$$

The French Yanguas-Gil found that there is solid carbon in the products of microwave discharge reforming of pure ethanol; deposition of carbon on the wall of the reactor can be avoided by adding water in the reforming process. The Portuguese Tatarova et al.[29] put forward the total reforming reaction formula of ethanol when studying the Ar/H_2O microwave discharge effects on ethanol reforming:

$$C_2H_5OH + \alpha H_2O \rightarrow (1+\alpha)CO + (3+\alpha)H_2 + (1-\alpha)C \tag{6.70}$$

$$C_2H_5OH + \alpha H_2O \leftrightarrow 2CO + 4H_2 + (\alpha - 1)H_2O \tag{6.71}$$

In addition, the reactor used in this study can keep the stream of materials at a high-speed vortex flow pattern in the discharge region, which has the following advantages in avoiding carbon deposition: (1) The stream of materials mixed well in the discharge area decreases the temperature gradient in the reactor, which weakens the diffusion of solid carbon on the wall of the tube[19]; (2) The gas flow rate can be up to subsonic and even supersonic, thereby weakening the attachment of solid carbon to the wall; (3) A thoroughly mixed stream of materials and high-concentration O and OH radicals in the discharge area also help the rapid oxidation of the solid carbon in the reaction zone.

6.1.5 Removal of NO_x in the Process of Reforming

Since nitrogen accounts for 78% of the volume fraction in the air, nitrogen oxides in the air plasma are very likely to be detected. When air with 48 m³/h is passed into the miniature non-thermal arc reactor, the concentration of nitrogen oxides produced in gas products is about 440 ppm. When the reforming conditions are O/C = 0.44, S/C = 1.28 and the net flow of ethanol is 0.10 g/s, the concentration of NO_x detected in the gas products is less than 20 ppm, and this means that the solution of water and ethanol inhibits and eliminates the effects on the generation of the nitrogen oxides.

In the air plasma, N_2 and O_2 are decomposed into a single molecule radical with the bombardment of high-energy electrons:

$$O_2 + e \rightarrow O + O + e \qquad (6.11)$$

$$N_2 + e \rightarrow N + N + e \qquad (6.12)$$

Then, the O radical, O_2 and N radicals generate NO through a complex reaction and then the oxidation of NO continues with the help of the oxidant O radical, and NO_2 is generated[30].

$$N + O \rightarrow NO \qquad (6.72)$$

$$N + O_2 \rightarrow NO + O \qquad (6.73)$$

$$NO + O \rightarrow NO_2 \qquad (6.74)$$

NO and NO_2 can be transformed into each other:

$$NO + O_3 \leftrightarrow NO_2 + O_2 \qquad (6.75)$$

$$NO + O_2 \leftrightarrow NO_2 + O \qquad (6.76)$$

In the non-thermal arc discharge process, the N_2 molecules require a high level of activation energy (9.8 eV) to be dissociated into N atoms. However, besides air, there is also a large number of water molecules and ethanol molecules involved in the reaction of the reforming process, which decreases the level of the average energy of the discharge area, thus the probability that N_2 molecules dissociate and generate N atom decreases, thereby limiting the generation of NO and NO_2. In addition, H_2O and ethanol molecules will also provide OH and H radicals in the discharge; soluble nitrate and nitrite generated through the reactions between these free radicals and NO and NO_2 are absorbed by water, fixed to the liquid phase[30–32]:

$$NO + OH \leftrightarrow HNO_2 \tag{6.77}$$

$$NO_2 + OH \leftrightarrow HNO_3 \tag{6.78}$$

$$HNO_2 + O \leftrightarrow HNO_3 \tag{6.79}$$

$$H_2O + O \leftrightarrow H_2O_2 \tag{6.80}$$

$$HNO_2 + H_2O_2 \leftrightarrow H_2O + HNO_3 \tag{6.81}$$

$$NO_2 + NO + H_2O \leftrightarrow 2HNO_2 \tag{6.82}$$

This process is also called plasma nitrogen fixation process.

Liquid phase and NO_2^- and NO_3^- in the liquid phase which were detected using high voltage pulse discharge processes a gas mixture of nitrogen/oxygen[31, 33]. Radicals and NO_x collide to generate HNO_2 and HNO_3, which can be transformed into NH_4NO_2 and NH_4NO_3 by NH_3[30]. Therefore, the non-thermal arc has broad application prospects in the removal of NO_x in the fumes or exhaust of the internal combustion engine.

6.2 Mechanism Analysis of the Plasma-Catalytic Reforming of Ethanol

As described in the first chapter, the methods of the catalytic process and the plasma process have their advantages and disadvantages. Currently, the catalyst technology is the most widely-used technology, which has the advantages of improving the conversion of ethanol and the selectivity of hydrogen. This process also has many shortcomings; for example, the catalyst is expensive and will be out of activity easily because of pollution or carbon deposition; this process has to be carried out under a high temperature, so the level of energy is high. And the size of the equipment and response time also limit its automotive applications. The plasma method is power-driven, and energy is only used to form the high-energy electrons with high reactivity. The background temperature in the reforming is low, so this method is characterized by a small size of the device, short response time, high reaction rate and energy efficiency; however, the conversion of ethanol and the selectivity of hydrogen in this process are far worse than the catalyst method. Therefore, it is hopeful to obtain the advantages of both methods by combining the catalyst method with the plasma method. If so, we can try our best to minimize energy consumption under the premise of high conversion and high selectivity of the product. Currently, the major applications of the plasma-catalyst combined process are the degradation of gaseous pollutants[34], the volatile organic compounds (VOCs)[35] and hydrogen production from CH_4 reforming[16, 36-37]. However, the knowledge regarding its application in the reforming of ethanol is very limited, which lets alone the mechanism. In last part of this work, the mechanism of the plasma-catalyst combination reforming will be discussed in detail with a look at the influence of plasma on the characteristics of the surface of the catalyst, the extension of the surface of the catalyst with the help of electron impact, free radicals, etc.

6.2.1 Related Mechanism of the Catalytic Reforming of Ethanol

Before analyzing the mechanism of plasma-catalyst reforming, getting the hang of the method of catalyst reforming will undoubtedly enhance the understanding of the mechanism of the combination reforming. The metal catalysts commonly used in the process of ethanol reforming are mainly non-precious metals such as Fe, Co, Ni, Cu and precious metals such as Ru, Rh, Pd, Pt, Ir, Laand Ce. Most of the above metals belong to Group VIII and Group IB of the periodic table of elements[26, 38].

Three kinds of non-precious metal catalysts, Ni, Cu and Co, were used in the research. Ni catalysts have strong activity in the destruction of C–C bonds, the C–H and C–O bond, so it is the most commonly used in the reforming of ethanol[39-40]. Co catalysts have high activity at low temperatures and can promote the breaking of the C–C bond and inhibit the machination of CO to reduce the amount of CH_4 generated[41]. It has been proved that Cu has strong activity in the WGS (water gas shift reaction) and the MSR (methane steam reforming reaction) but poor activity in the destruction of the C–C bond, so it is often used as an added ingredient with other active metals in the catalytic reforming of ethanol[39].

Take the Ni catalyst for example: literature[42] simplifies the process of the degradation of alcohol. In connection with the literature[39, 42-45], the related reactions of ethanol in the catalytic auto-thermal reforming are as follow:

Water is absorbed by the surface of the catalyst and decomposed into adsorbed H and OH. These two adsorbed materials play a role in the catalytic reforming of ethanol:

6 Mechanism for the Plasma Reforming of Ethanol

$$H_2O + (a) \leftrightarrow H_2O(a) \qquad (6.83)$$

$$H_2O(a) + (a) \leftrightarrow H(a) + OH(a) \qquad (6.84)$$

The reactions after the ethanol is adsorbed by the surface of the catalyst are as follow:

1. Dehydrogenation and degradation reactions of ethanol

This is the first reaction that produces hydrogen:
Total formula:

$$CH_3CH_2OH \leftrightarrow CH_3CHO + H_2 \qquad (6.56)$$

Elementary reaction:

$$CH_3CH_2OH + (a) \leftrightarrow CH_3CH_2OH(a) \qquad (6.85)$$

$$CH_3CH_2OH(a) + (a) \leftrightarrow CH_3CHOH(a) + H(a) \qquad (6.86)$$

$$CH_3CHOH(a) + (a) \leftrightarrow CH_3CHO(a) + H(a) \qquad (6.87)$$

$$2H(a) \leftrightarrow H_2 + 2(a) \qquad (6.88)$$

2. Degradation reaction of acetaldehyde

The C=C bond of the acetaldehyde breaks and CH_4, CO are generated, but the Cu catalyst has a poor ability to break the C=C bond, so the main degradation products are CH_3CHO or C_2H_4:
Total formula:

$$CH_3CHO \leftrightarrow CH_4 + CO \qquad (6.89)$$

Elementary reaction:

$$CH_3CHO(a) \leftrightarrow CH_3CHO + (a) \qquad (6.90)$$

$$CH_3CHO(a) + (a) \leftrightarrow CH_3(a) + CHO(a) \qquad (6.91)$$

$$CH_3CHO(a) + (a) \leftrightarrow CH_4(a) + CO(a) \qquad (6.92)$$

$$\text{CHO(a)} + \text{(a)} \leftrightarrow \text{CO(a)} + \text{H(a)} \tag{6.93}$$

$$\text{CH}_4(\text{a}) + \text{(a)} \leftrightarrow \text{CH}_3(\text{a}) + \text{H(a)} \tag{6.94}$$

$$\text{CH}_3(\text{a}) + \text{H(a)} \leftrightarrow \text{CH}_4(\text{a}) + \text{(a)} \tag{6.95}$$

$$\text{CO(a)} \leftrightarrow \text{CO} + \text{(a)} \tag{6.96}$$

$$2\text{H(a)} \leftrightarrow \text{H}_2 + 2(\text{a}) \tag{6.88}$$

3. The water-gas shift

H_2O molecules are used to oxidize CO into CO_2 and additional hydrogen is produced in this reaction. A low temperature (<400 °C) favors the proceeding of this reaction.

Total formula:

$$\text{H}_2\text{O} + \text{CO} \leftrightarrow \text{H}_2 + \text{CO}_2 \tag{6.97}$$

Elementary reaction:

$$\text{H}_2\text{O} + \text{(a)} \leftrightarrow \text{H}_2\text{O(a)} \tag{6.83}$$

$$\text{H}_2\text{O(a)} + \text{(a)} \leftrightarrow \text{H(a)} + \text{OH(a)} \tag{6.84}$$

$$\text{CO(a)} + \text{OH(a)} \leftrightarrow \text{COOH(a)} + \text{(a)} \tag{6.98}$$

$$\text{COOH(a)} + \text{(a)} \leftrightarrow \text{CO}_2(\text{a}) + \text{H(a)} \tag{6.99}$$

$$2\text{H(a)} \leftrightarrow \text{H}_2 + 2(\text{a}) \tag{6.88}$$

4. The reforming of methane steam

This reaction can abstract the hydrogen atoms of the methane and increase the selectivity of hydrogen:

Total formula:

$$\text{CH}_4 + \text{H}_2\text{O} \leftrightarrow \text{CO} + 3\text{H}_2 \tag{6.100}$$

Elementary reaction:

$$\text{CH}_4(\text{a}) \leftrightarrow \text{CH}_4 + \text{(a)} \tag{6.101}$$

$$\text{H}_2\text{O} + \text{(a)} \leftrightarrow \text{H}_2\text{O(a)} \tag{6.83}$$

$$\text{H}_2\text{O(a)} + \text{(a)} \leftrightarrow \text{H(a)} + \text{OH(a)} \tag{6.84}$$

6 Mechanism for the Plasma Reforming of Ethanol

$$CH_4(a) + (a) \leftrightarrow CH_3(a) + H(a) \tag{6.94}$$

$$CH_3(a) + OH(a) \leftrightarrow CO(a) + 2H_2 + (a) \tag{6.102}$$

$$CO(a) \leftrightarrow CO + (a) \tag{6.96}$$

$$2H(a) \leftrightarrow H_2 + 2(a) \tag{6.88}$$

5. The dehydration reaction of ethanol

The dehydration reaction of ethanol will reduce the hydrogen production and the ethylene generated in the reaction is the main cause of carbon deposition. The ideal process of the catalysis of ethanol needs to break the $C=C$ bonds of the reactants and the intermediate:

Total formula:

$$CH_3CH_2OH \leftrightarrow C_2H_4 + H_2O \tag{6.57}$$

Elementary reaction:

$$CH_3CH_2OH + (a) \leftrightarrow CH_3CH_2OH(a) \tag{6.85}$$

$$CH_3CH_2OH(a) + (a) \leftrightarrow C_2H_4(a) + H_2O(a) \tag{6.103}$$

$$C_2H_4(a) \rightarrow \text{polymer} \rightarrow \text{deposition of carbon} \tag{6.58}$$

6. Reactions that oxygen is involved in

Oxygen plays the role of dehydrogenation in the reaction and oxidizes some of the carbon deposition and CO into CO_2:

$$CH_3CH_2OH + \frac{1}{2}O_2 \leftrightarrow CH_3CHO + H_2O \tag{6.104}$$

$$CH_4 + \frac{1}{2}O_2 \leftrightarrow CO + 2H_2 \tag{6.105}$$

$$C + \frac{1}{2}O_2 \leftrightarrow CO \tag{6.106}$$

$$CO + \frac{1}{2}O_2 \leftrightarrow CO_2 \tag{6.107}$$

$$H_2 + \frac{1}{2}O_2 \leftrightarrow H_2O \tag{6.108}$$

$$O_2 + 2(a) \leftrightarrow 2O(a) \qquad (6.109)$$

$$CH_3CH_2OH(a) + O(a) \leftrightarrow CH_3CH_2O(a) + OH(a) \qquad (6.110)$$

$$CH_3CH_2O(a) + O(a) \leftrightarrow CH_3CHO(a) + OH(a) \qquad (6.111)$$

$$CH_3CHO(a) + O(a) \leftrightarrow CH_3CO(a) + OH(a) \qquad (6.112)$$

$$C(a) + O(a) \leftrightarrow CO(a) + (a) \qquad (6.113)$$

$$CO(a) + O(a) \leftrightarrow CO_2(a) + (a) \qquad (6.114)$$

$$H(a) + O(a) \leftrightarrow OH(a) + (a) \qquad (6.115)$$

$$OH(a) + O(a) \leftrightarrow H_2O(a) + (a) \qquad (6.116)$$

In addition, oxygen is used as a catalyst carrier in the research. The mechanical strength and chemical resistance of this carrier are very strong, so it is widely used in the catalytic processes[46]. Reactions that will occur on the γ-Al_2O_3 are as follow:

$$C_2H_5OH \rightarrow C_2H_4 + H_2O \qquad (6.117)$$

Besides, γ-Al_2O_3 can degrade the ozone and generate an active O atom[47]:

$$O_3 + (a) \rightarrow O_2 + O(a) \qquad (6.118)$$

6.2.2 Effects of Plasma on the Surface Characteristics of the Catalyst

There are two main combination forms of plasma and catalyst: the single-stage system and the secondary system[13]. In the single-stage system, the plasma discharge region covers a part of or the complete surface of the catalyst, so plasma will unavoidably have an impact on the characteristics of that surface. Besides, the reaction of the plasma and the reaction of the catalyst happen in the space, so the mechanism is more complex; in the secondary system, there is no direct interaction between the plasma and the catalyst, so the effect of reforming can be regarded as the simple sum of the two processes[13]. The coupled reforming system used in this study was a single-stage system, whose catalyst bed was in the discharge region of the plasma. It is worth noting that in much of the literature, the dielectric barrier discharge (DBD) was used in the single-stage system[7, 48–49], and the catalyst was located directly between the two electrodes, so the reaction of plasma and the activation reaction of the catalyst were carried out almost in the same space. In this study, the catalyst bed was located at the rear of the gap of the electrode pair. However, the plasma extended to the downstream of the discharge region during the boost of the high-speed airflow, thus the plasma was able to affect the catalytic bed as well.

The effects on the characteristics of the surface of the catalyst can be divided into two categories: one is that the non-thermal plasma can change the surface structure of the catalyst, thus optimizing the size of the particles and the dispersion of the metal, and this process is irreversible. The other one is the activation of the plasma on the catalyst surface at a low temperature, which only exists in the discharge process.

Currently, the plasma technology, especially the non-thermal plasma technology, has been widely used in the surface modification of the materials[50-51], which is usually used to change the physico-chemical properties of the materials and is not involved in the later material reaction. Hence, the reconstruction effects of non-thermal plasma on the physicochemical properties of the materials can be analyzed separately. In the process of the reforming catalyst, increasing the specific surface area and reducing the size of the particles of the active metal usually means an increase in the ability to produce hydrogen when the other conditions are the same[52]. In the non-thermal plasma environment, high-density energetic electrons and living radicals bombard the catalyst. As a result, catalyst particles with a larger surface size break, and even smash. Therefore, the surface size of the catalyst diminishes significantly[53-54]. In addition, the surface particles of the material capture and gather a large number of electrons in a very short time under the action of electron flow that the plasma generates and then sheath is formed on the surface of the particle. There is a strong Coulomb repulsion between this electron sheath and the electron flow on the surface of the bombarding material, so large particles of an irregular shape are easier to crack and separate in the gap, thereby reducing the size of the particles and raising the specific surface area[55]. Besides, the Coulomb repulsion between the particle sheath also facilitates the uniform distribution of the particle spacing, thereby improving the dispersion of active sites on the surface of the catalyst[53, 56-57]. In addition, electron sheath can attract deeper active metal ions deeply inside the catalyst voids, making the metal ions migrate to the surface of the catalyst in the electric field, and thus improving the utilization of the active metal in the surface[58].

In summary, a catalyst can provide more effective active sites after being processed by the plasma, and the reactants are absorbed more easily on the catalyst surface. The reaction happens more easily, thereby improving the catalytic activity. The Ni/SiO_2 catalyst was prepared by an atmospheric pressure plasma jet[56], and it was found that the specific surface area of the catalyst and the dispersion of Ni can be improved by using plasma as an auxiliary and having the activation temperature of the catalyst be 50 °C, lower than that of the traditional method. Glow plasma at room temperature was used to process NiO which photo catalyst often supported and found that plasma can control the interface of metal-carrier and metal particles with small and uniform size which can be obtained[59]. The catalytic activity of the prepared NiO/ Ta_2O_5 is 1.7 times that of conventional processing methods. Haghighi et al. of Iran[55] found that plasma processing will enable a higher activity of the catalyst at low temperatures and the higher conversion of the materialism the process of preparing the Ni/Al_2O_3–MgO nano catalyst by using plasma as an auxiliary. In addition, plasma can produce new active sites on the catalyst, thereby widening the pathways for the reactions[60–61].

A reduction in the size of surface particles and improvement in dispersion can also enhance the interaction between the active metal components and the carrier, thereby improving the stability of the catalyst by inhibiting sintering or coking by suppressing sintering or the phenomenon[56, 62–63]. Factors leading to deactivation of the catalyst include carbon deposition and sintering; therefore, the stability of the catalyst in the operation is related to the factors of metal loading, the size of the grain and the interaction between the active ingredient and the carrier and so on[64]. So a strong interaction between the metal particles and the carrier helps to avoid aggregation of metal particles due to their migration[64]. Take the Ni/Al_2O_3 catalyst as an example, the Ni particle has a smaller size after the processing of the plasma and it can migrate into an Al_2O_3 carrier and generate $NiAl_2O_4$ more easily. It has been found that Ni has two major forms, $NiAl_2O_4$ and microcrystalline NiO, in the Ni/γ-Al_2O_3 catalysts[63], in which the former one has high dispersion and weak mobility since the active metal interacts strongly with the carrier, and therefore it has high activity and resistance; the latter one has strong mobility and its activity is easy to reduce because of carbon deposition and sintering. It has been found that the Ni particles of the nano-catalyst Ni/Al_2O_3–MgO after being processed by glow discharge has high dispersion and low particle size, thereby reducing the structural defects of the catalyst surface and inhibiting the diffusion of carbon deposition in the active site[55]. The Cu/Al_2O_3 catalyst has a similar phenomenon. In addition, the two opposing processes of generation of solid carbon and gasification of the carbon exist at the same time in the process of reforming, the phenomenon of carbon deposition occurs when the rate of formation of solid carbon is significantly higher than the rate of the gasification of carbon, and the increase of the specific surface area strengthens the adsorption capacity of the catalyst to absorb oxidative O free radical and CO_2, increasing the vaporization rate of solid carbon and thus weakening or eliminating the carbon deposition phenomenon[52, 55].

It is worth mentioning that the strong electric field that the plasma generates in the process of discharge also affects the activity of the catalyst. The strong electric field generated by the discharge generates a dielectric polarization on the catalyst surface, making the active metal absorb electrons and promote the catalyzed reaction[65]. The electric field coupling with Pd, Rh and Pt catalysts at low temperatures (423–573 K) was used to reform ethanol and it was found that at 473 K, the electric field can make the conversion rate of ethanol increase from 3.7% to 71.2% when using Rh/CeO_2

catalysts at low temperatures. This is equivalent to the situation in which the catalytic temperature drops to at least 150 K. Theoretical calculations also show that after adding the electric field, activation energy of all reactions in the reforming of ethanol decrease by 20%–90%, which is more obvious in the water gas shift reaction and the methane steam reforming reaction.

6.2.3 The Surface Reaction of the Electronic/Radical-Catalyst

In addition to increasing the number of active sites and the activity of the catalyst, the discharging reforming process can also play a series of synergies affects with the catalyst, for example: (1) the heat of the plasma reforming process can heat the catalyst; (2) the active particles generated can degrade the substrate molecules that are involved in the catalytic surface reaction; (3) the active radicals generated by the plasma are involved in the catalyst surface reaction directly in the forms of homogeneous reaction and heterogeneous reaction; (4) selectivity of the catalysts helps to increase the selectivity of the objective product.

1. The heating effect of plasma reforming on the catalyst

In the experiment, the air flow rate is 0.48 m^3/h, and it was found that the temperature of the air flow increased from 25 °C to 42–46 °C, which suggests that the stream of ethanol/water/air mixture cannot be heated to a relatively high temperature only by electricity. However, the temperature of the reaction zone in the oxidation steam reforming of ethanol was up to 120–210 °C, which shows that the temperature of the reaction zone came mainly from the heat of the degradation of ethanol. However, the temperature was much lower than that of the conventional catalytic processes (500–800 °C), showing that the processing of plasma can significantly reduce the temperature required for catalytic reforming, which will help to use a more flexible structure and materials to make the reforming reactor. It is worth mentioning that, in the catalytic auto-thermal reforming of ethanol, although the total reaction is a thermodynamically neutral formula, we still need to provide an external heat source to heat the reactor in the actual operation. According to the theoretical calculations, without considering heat loss, when the molar ratio of Ethanol/H_2O/O_2 is 1.0:2.0:0.5 and the net flow of ethanol is 0.10 g/s, the thermal power required to heat the raw materials to 500–800 °C is 533–724 W, which is equivalent to 19%–26% of the calorific value of ethanol (rate of water, and can be up to 3–6 in the actual reforming, and the required thermal power is equivalent to 24%–49% of the calorific value of ethanol), which is 1–2 orders of magnitude higher than that of non-thermal arc plasma-catalyst combination reforming under the same feeding conditions (12–15 W).

2. Excitation and degradation effects of the substrate of the plasma reaction

The gaseous reaction substrate reaches the surface of the catalyst in the form of the ground state in the catalytic reaction and the catalytic processes are all completed at or near the surface of the catalyst[13], when constructing the mechanism of the Cu/ZnO/Al$_2$O$_3$ catalyst for the steam reforming of methanol. It has been suggested that there is at least one catalytic site which is capable of adsorbing CH$_3$OH, CO$_2$, CO and H$_2$O and groups of atoms such as HCOO$^-$, CH$_3$O$^-$, HCO$^-$, etc., and thus establishing that ethanol and methanol have similar structures[66]. Therefore, it can be speculated that there is competitive adsorption between some molecules or atoms (groups) on the surface of the catalyst in the conventional catalytic reforming of ethanol, so the activity of the catalyst is restricted. In addition, the reaction substrates need a discharge region before reaching the catalytic bed, and some of the molecules are excited to the excited-state particles or decomposed into molecular fragments in the role of high-energy electrons, and active radicals affect the type of the leading reaction of high-energy electron collisions in the plasma reaction. When the average energy level of electrons is higher, dissociation and ionization reactions of molecules are dominant; conversely the vibrational excitation of molecules is dominant[5, 17]. In the non-thermal arc discharge, the upstream of the discharge region is the thermal equilibrium state, and the average energy level of the electrons can be 5–8 eV or more; due to heat loss, the average energy level of the electrons drops to 1–3 eV in the downstream of the discharge region. The energy threshold required to excite the molecules into a state of rotational excitation, a state of vibrational excitation and a state of electronic excitation by electrons is 0.01–0.1 eV, 0.1–1.0 eV and 1.0–10 eV specifically[13], so excited-state particles that reach the surface of the catalyst are mainly in a state of rotational excitation and a state of vibrational excitation, and a small amount are in a state of electronic excitation, wherein the energy level of the molecules in the state of rotational excitation is so low that they have no significant promotion for the reforming, so the major excited molecules that participate in the synergy are vibrational molecules. Compared with the ground state, vibrationally-excited particles have a lower activation energy in the adsorption and reaction process, which in the process of reforming (1) can improve the activity and the utilization of the active sites, thus speeding up the chemical adsorption dissociation of the reaction substrates; (2) can maintain the number of active sites by increasing the vaporization rate of the carbon deposition, thereby increasing the stability and efficiency of the catalyst. These effects are both conducive to accelerating the rate-controlled reaction and overall reaction in the catalyst reaction[13, 67]. Beck et al.[68–69] obtained a specific vibrationally-excited molecule beam CH$_4$ through resonant narrowband laser irradiation and explored its effects on gas-surface reaction kinetics. Their experiment showed that vibrational excitation can significantly improve the sticking coefficient of gaseous molecules. In addition, the Japanese Nozaki et al.[16] proposed, in the catalytic-DBD discharge combination reforming of methane, that the vibrational excitation-state CH_4 generated by high-energy electrons can promote the dissociative chemisorptions of CH_4 on the surface of the Ni catalyst, and excited-state H_2O can also remove the carbon deposition of the catalyst so as to maintain the number of active sites of the catalyst.

Degradation of ethanol generates a variety of intermediate products in the non-thermal arc discharge. Section 6.2.1 provides the general method of the reforming of ethanol by the traditional catalytic process. The flows of raw materials that reach the surface of the catalyst are ground-state molecules such as C_2H_5OH, H_2O, O_2 and N_2 in the traditional catalytic method; later there is competition on adsorption and reaction on the catalyst active sites between raw materials such as C_2H_5OH and adsorbed molecules and molecular fragments of different degrees of degradation (that is CH_3CHOH, CH_3HCO, HCO, CH_4 and CH_3, etc.). The rate of reforming is restricted by rate-controlling, therefore, the processing capacity of the catalyst is limited. The dehydrogenation of ethanol of the copper-based catalyst was examined and it was found that the reforming products were acetaldehyde and H_2 at the temperature of 200–400 °C while the main reforming product of the CuO/Al_2O_3 catalyst was mainly C_2H_4 above the temperature of 350 °C[70]. There was no generation of H_2 and the degree of fracture of C–C bonds in ethanol was extremely low. Cu/γ-Al_2O_3 was utilized to reform ethanol and it was found that although there was a certain amount of CH_4 in the gas products, there was still 59% of the C element existing in the form of C_2 particles (mainly C_2H_4)[71]. It was calculated according to BEP (Brønsted-Evans-Polanyi) relations and it was ascertained that the C–C bond cleavage ability of the Cu catalyst in the degradation of ethanol is very weak, and its rate-controlling step is the decomposition of the intermediate product CH_3CHO[72]. Moreover, the reaction substrate that has the lowest activation energy of the reaction of C–C bond cleavage is CH_3CO. Non-thermal arc discharge can weaken the restrictive state of the rate-controlling step by decomposing the CH_3CHO molecule; in addition, a certain amount of CH_3CO was generated in the discharge degradation, thereby promoting C–C bond cleavage and significantly improving the conversion rate of ethanol. When the catalysts were Ni and Co, non-thermal arc discharge was also able to degrade some of the ethanol and its molecular fragments previously, thereby making the molecule and the molecular debris particles, between which there was competition of adsorption and reaction, migrating to the small molecular weight particles.

3. Participation of radicals on the catalyst surface reaction

The convergent-divergent tube structure was used in the research regarding Plasma-catalyst combination reforming. The cross-sectional area of the channel shrank rapidly along the convergent direction of the convergent-divergent tube, and material flow reached the subsonic and even the supersonic scale in the throat quickly. Although the cross-sectional area of the channel increased along the divergent direction later, the material flow sped up in the Laval nozzle effect and the air pressure plummets. At this time, existence time, dimensions and activity of high-energy electrons improved so that they were able to reach the surface of the catalyst and be involved in the synergistic effect[17]. It is worth noting that due to heat loss and energy exchange, the energy level of electrons in the downstream of the discharge area was reduced significantly, so the catalyst surface reaction was primarily participated in by the active free radicals and the radicals promoted the conversion and the degradation of the reaction substrate of the catalyst surface reaction. According to the literature, single-stage plasma-catalyst combination reforming often uses the configuration that the discharge region is coincident with in the catalytic bed, so the plasma reaction and catalytic reaction occur in the same space and time, and there is competition among the materials of the reforming and the intermediate products of different degrees of degradation. Kim[7] from South Korea found that in the discharge-catalytic system, steam reforms of methanol and that it will reduce the conversion of methanol when increasing the discharge voltage from 3 to 4 kV at the temperature of 250 °C; and this phenomenon does not appear at the temperature of 180 °C. He believed that the number of free radicals such as CH_3, H and CH_3O and so on increases rapidly at high voltage and temperature, which increase their competition for the catalyst adsorption sites with the methanol, resulting in a negative impact on the direct degradation of methanol[7]. In this study, the concentration of the materials such as ethanol diminished significantly because of their degradation in the upstream of the discharge region, which had attenuation effects on the competition for the adsorption sites.

There are two mechanisms for the catalyst surface reaction: the Langmuir-Hinshelwood reaction and the Eley-Rideal reaction. The former refers to reactions between two chemical adsorbed reactants; the latter is a reaction between chemical adsorbed reactants and gaseous reactants[13]. As mentioned earlier, non-thermal arc discharge generates a large amount of active free radicals and these active free radicals play an important role in the degradation of the substrate and in the distribution of the products. In terms of the Langmuir-Hinshelwood reaction, since the active free radicals have a higher energy level and are reactive, their sticking coefficient is higher and they are easier to be captured by the surface of the catalyst and participate in the surface reaction[13]. In terms of the Eley-Rideal reaction, the body in the conventional surface catalysis reaction is the adsorbed molecules/molecular fragments and the ground-state gaseous reactants, while active free radicals such as H, O, OH and CH_3 and so on produced through non-thermal arc discharge can react with adsorbed reactants directly, thereby widening the pathway of the surface catalyzed reaction[7]. In summary, the high-energy electrons and active free radicals generated in the non-thermal arc plasma can strengthen the catalyst surface reaction from the two mechanisms.

4. Improvement of plasma reformate by the reaction selectivity of the catalyst

In the non-thermal plasma environment, high-energy electrons can excite and/

or degrade gas molecules rapidly at low background temperatures, but since the distribution of the overall energy level of high-energy electrons is difficult to control and the high-energy electrons with the same energy level are non-selective during the initiation of the reaction, so the selectivity of the target product is always not high. Introducing a catalyst into the plasma reforming system helps to converse the non-target products so as to improve the yield of the target product and the energy efficiency of the reforming. It is worth mentioning that the Al_2O_3 carrier at the rear of the discharge area can absorb and decompose O_3 molecules and generate adsorbed O atoms, which plays an important role in improving the participation of the O_2 molecules in the reforming reaction and it also helps the oxidation and removal of carbon deposition on the catalyst surface, thus ensuring the stability and activity of the catalyst[47].

6.3 Comparison Between Plasma Reforming and Plasma-Catalyst Reforming

Chapter 4 shows that adding a non-precious metal catalyst to the non-thermal arc reforming system can improve the conversion of ethanol, the selectivity of the product and the unit energy consumption of hydrogen production. To highlight the advantages of the plasma-catalyst combination reforming process, we will compare this process with the separate non-thermal arc reforming process regarding three aspects: the reforming mechanism, the reforming performance and the reforming energy:

1. Investigation regarding the aspect of the reforming mechanism

In the separate non-thermal arc reforming process, the main promoters of the reaction are high-energy electrons and active free radicals. These two active particles can stimulate and/or dissociate the raw material of the reforming and generate various fragments of molecules in the discharge region where the average energy level is very high. Later, complex reactions of the molecular fragments occur and all kinds of products are generated and the main gas products include H_2, CO, CO_2 and CH_4.

The reaction zone can be divided into three sections in the non-thermal arc-catalytic combination reforming: the simple discharge area, the region of the discharge-catalyst layer and the region of the back catalytic layer. The reforming mechanism of the simple discharge region is substantially the same as that of the separate non-thermal arc reforming process; the difference is that the length of this region is shorter than that of the discharge region of the separate non-thermal arc reforming process; in the region of the discharge-catalyst layer, the average energy of the particles is low because they are at the rear of the discharge area. Therefore, the main participants in the discharge reforming reaction are radicals which have a long life and can continue to participate in the dissociation reaction and high-energy electrons mainly by playing the role of excitation. Due to the low-temperature characteristics of the micro non-thermal arc, the thermal effect of the catalytic bed is mainly provided by the discharge degradation of the ethanol in the discharge area and it maintains at a low level of temperature (lower than 220 °C). In this temperature range, conventional catalysts have substantially no catalytic activity, but because of the high adhesion, the reactivity of the radicals and the modification, as well as the activation effects on the catalyst surface by high-energy particles, the catalyst still has a certain activity at low temperatures. The final products are steadily distributed in the region of the discharge-catalytic layer and later outflow via the back catalytic layer.

2. Investigation regarding the aspect of the reforming performance

In terms of the conversion rate of ethanol, the conversion rate of the combination reforming can be above 80%, while that of the separate non-thermal reforming process can only be about 50%, which is because the non-thermal arc excites the activity at low temperatures of the catalysis. Compared with the simple non-thermal arc reforming process, combination reforming has a higher selectivity of H_2 and a lower selectivity of CH_4, which indicates that the catalytic bed has a transformation capacity of CH_4. Overall, in the combination reforming, the catalyst bed has a complementary role in conversing the reaction substrate and changing the component of the products.

3. Investigation regarding the aspect of the reforming energy

The level of the electric power of the non-thermal arc reforming system and the combination reforming system is very low (12–15 W), so the heat that makes the reaction system reach a steady temperature and makes the catalyst get the activity is mainly from the degradation of ethanol. Although the degree of degradation of ethanol is higher in the combination reforming system, the reaction temperature of the combination reforming system is higher than that of the non-thermal arc reforming system. So, regarding the former, more of the chemical energy released is used to heat the system, so it cannot significantly increase the energy efficiency. However, under similar electric power, the combination reforming system can obtain a higher conversion rate of ethanol and higher selectivity of hydrogen at the same time, thereby obtaining higher output of hydrogen and lower energy consumption of the per unit production of hydrogen.

Fig. 6.3 shows the energy requirement for heating 1 mol of H_2O, 1 mol of ethanol, and air containing 1 mol of O_2 to different temperatures. Transformation points of water and ethanol are 351.3 and 373.15 K respectively, and heating these

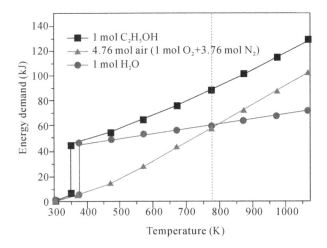

Fig. 6.3 The energy requirement for heating 1 mol of H_2O, 1 mol of C_2H_5OH and air containing 1 mol of O_2 from ambient temperature (298.15 K) to higher temperatures (298–1073 K), respectively (without consideration of heat losses)

two substances across their transformation points would consume a lot of energy to vaporize. Therefore, if we can make the high-energy electrons obtain a higher density and energy level at a lower background temperature (such as below 351.3 K), then the plasma is expected to significantly improve the energy efficiency of the reforming system and the non-metallic catalysts may be given stronger catalytic activity at a low temperature. Future research on plasmas on a smaller scale, in other words, the exploration of micro-plasma, will provide a possibility for putting this idea into practice.

6.4 Summary

This chapter is intended to analyze the microscopic mechanism of non-thermal arc plasma ethanol auto-thermal reforming and the plasma-catalytic auto-thermal reforming of ethanol in detail. The main contents of this chapter include the electron-molecule collision reaction and the radical reaction in non-thermal arc reforming, the effects on the physical characteristics of the surface of the catalyst and the surface catalytic reaction by discharge of non-thermal as well as argumentation on the feasibility and the effectiveness of combining the miniature non-thermal arc and the catalytic bed to improve the effectiveness of reforming. As described above, in the process of plasma-catalyst combination reforming of ethanol, synergy between the plasma and the catalyst, transportation and transformation of the materials in the reaction region are shown in Figs. 6.4 and 6.5 respectively. The main conclusions of this chapter are as follows:

(1) The most important reactions in the non-thermal arc reforming of ethanol are the electron-molecule collision and the radical reaction, wherein the high-energy electron collision reaction makes the heavy particles excited and/or dissociate into a variety of molecular fragments and free radicals. There are a lot of OH, H, O and CH_3 free radicals existing in the discharge environment, and their interaction with other particles has a significant impact on the types and distribution of the reformation.

(2) There is competition between hydrogen-producing reactions and methane-producing reactions in non-thermal arc reforming, which limits the selectivity of hydrogen reforming. It can improve the efficiency of hydrogen production when adding the catalyst into the reforming system and using the catalyst to degrade the methane.

(3) In the non-thermal arc-catalytic reforming system, the discharge area covers the catalyst surface, so discharge can change the surface physicochemical properties

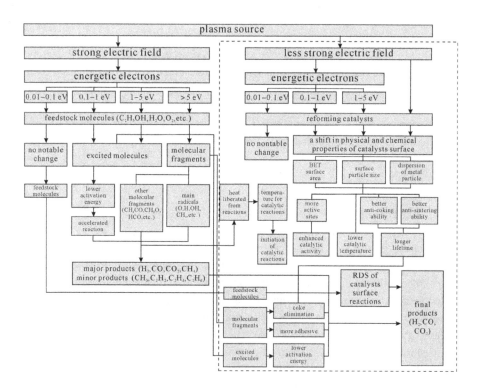

Fig. 6.4 Synergistic effects between plasma and catalyst during ethanol reforming processes

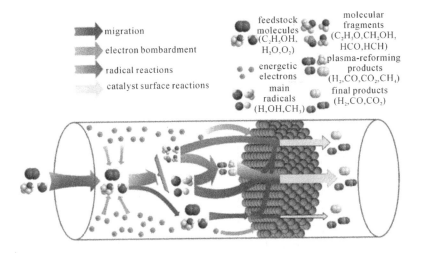

Fig. 6.5 Synergistic effects between plasma and catalyst during ethanol reforming processes

of the catalyst, thereby obtaining the particles with a smaller size and a better dispersion of the metal. In addition, active particles (high-energy electrons, free radicals) generated by the plasma have activation on the active site of the catalyst at a low temperature.

(4) In the plasma-catalytic reforming system, reaction heat released in the non-thermal arc discharge decomposition of ethanol can help the catalytic bed reach the desired temperature, and the intermediate produced through the degradation of alcohol in the discharge is beneficial for the catalytic reactions to break the limitation of the rate-controlling step. In addition, active free radicals can strengthen the surface reactions of the catalyst of two reaction mechanisms, i.e. the Langmuir-Hinshelwood reaction and the Eley-Rideal reaction.

References

[1] Huang DW. Design and application of miniaturized nonthermal arc plasma for hydrogen generation from ethanol reforming. Sun Yat-sen University. 2014.

[2] Levko DS, Tsymbalyuk AN, Shchedrin AI. Plasma kinetics of ethanol conversion in a glow discharge. Plasma Phys Rep. 2012; 38(11): 913–921.

[3] Itikawa Y, Mason N. Cross sections for electron collisions with water molecules. J Phys Chem Ref Data. 2005; 34(1): 1–22.

[4] Raizer YP. Gas discharge physics. J Atmos Sol-Terr Phy. 1993; 55(10): 1487.

[5] Levko D, Shchedrin A, Chernyak V, Olszewski S, Nedybaliuk O. Plasma kinetics in ethanol/water/air mixture in a 'tornado'-type electrical discharge. J Phys D Appl

Phys. 2011; 44(14): 145206–145218.
[6] Shirai T, Tabata T, Tawara H, Itikawa Y. Analytic cross sections for electron collisions with hydrocarbons: CH_4, C_2H_6, C_2H_4, C_2H_2, C_3H_8, and C_3H_6. Atom Data Nucl Data. 2002; 80(2): 147–204.
[7] Kim T, Jo S, Song YH, Lee DH. Synergetic mechanism of methanol-steam reforming reaction in a catalytic reactor with electric discharges. Appl Energ. 2014; 113(1): 1692–1699.
[8] Wang WJ, Zhu CY, Cao YY. DFT study on pathways of steam reforming of ethanol under cold plasma conditions for hydrogen generation. Int J Hydrogen Energ. 2010; 35(5): 1951–1956.
[9] Cao XL. Study on ethanol conversion with dielectric-barrier discharge. TianJin University. 2009.
[10] Futamura S, Kabashima H. Effects of reactor type and voltage properties in methanol reforming with nonthermal plasma. Ieee T Ind Appl. 2004; 40(6): 1459–1466.
[11] Yan ZC, Chen L, Wang HL. Hydrogen generation by glow discharge plasma electrolysis of ethanol solutions. J Phys D Appl Phys. 2008; 41(15): 1525–1528.
[12] Zhou ZP. Reforming of methane for hydrogen production via non-equilibrium plasma. University of Science and Technology of China. 2012.
[13] Chen HL, Lee HM, Chen SH, Chao Y, Chang MB. Review of plasma catalysis on hydrocarbon reforming for hydrogen production—Interaction, integration, and prospects. Appl Catal B-Environ. 2008; 85(1): 1–9.
[14] Nozaki T, Muto N, Kado S, Okazaki K. Dissociation of vibrationally excited methane on Ni catalyst—Part 1. Application to methane steam reforming. Catal Today. 2004; 89(1–2): 57–65.
[15] Van de Meerakker SYT, Vanhaecke N, van der Loo MPJ, Groenenboom GC, Meijer G. Direct measurement of the radiative lifetime of vibrationally excited OH radicals. Phys Rev Lett. 2005; 95(1): 13003–13200.
[16] Nozaki T, Tsukijihara H, Fukui W, Okazaki K. Kinetic analysis of the catalyst and nonthermal plasma hybrid reaction for methane steam reforming. Energ Fuel. 2007; 21(5): 2525–2530.
[17] Hammer T, Kappes T, Baldauf M. Plasma catalytic hybrid processes: Gas discharge initiation and plasma activation of catalytic processes. Catal Today. 2004; 89(1–2): 5–14.
[18] Baulch DL, Bowman CT, Cobos CJ, Cox RA, Just T, Kerr JA, Frank P, Hayman G, Murrells T, Pilling MJ, Troe J, Walker RW, Warnatz J. Evaluated kinetic data for combustion modeling: Supplement II. J Phys Chem Ref Data. 2005; 34(3): 757–1397.
[19] Tsyganov D, Bundaleska N, Tatarova E, Ferreira CM. Ethanol reforming into hydrogenrich gas applying microwave 'tornado'-type plasma. Int J Hydrogen Energ. 2013; 38(34): 14512–14530.
[20] Marinov N, Malte P. Ethylene oxidation in a well-stirred reactor. Int J Chem Kinet.

1995; 27(10): 957–986.
[21] Marinov N, Pitz W, Westbrook C, Vincitore A, Castaldi M, Senkan S. Aromatic and polycyclic aromatic hydrocarbon formation in a laminar premixed n-butane flame. Combust Flame. 1998; 114(1–2): 192–213.
[22] Marinov NM. A detailed chemical kinetic model for high temperature ethanol oxidation. Int J Chem Kinet. 1999; 31(31): 183–220.
[23] GRI-Mech Version 3.0 7/30/99 http://www.gri.org.
[24] Tsang W, Hampson R. Chemical kinetic data base for combustion chemistry. Part I. Methane and related compounds. J Phys Chem Ref Data. 1986; 15(3): 1087–1279.
[25] Golden DM. The Reaction OH + C_2H_4: An example of rotational channel switching. J Phys Chem A. 2012; 116(17): 4259–4266.
[26] Mattos LV, Jacobs G, Davis BH, Noronha FB. Production of hydrogen from ethanol: Review of reaction mechanism and catalyst deactivation. Chem Rev. 2012; 112(7): 4094–4123.
[27] Li HX. Design, characteristics and application of non-thermal Laval nozzle arc plasma reactor for hydrogen production of bio-ethanol reforming. Sun Yat-sen University. 2012.
[28] Wang YF, Tsai CH, Chang WY, Kuo YM. Methane steam reforming for producing hydrogen in an atmospheric-pressure microwave plasma reactor. Int J Hydrogen Energ. 2010; 35 (1): 135–140.
[29] Bundaleska N, Tsyganov D, Tatarova E, Dias FM, Ferreira CM. Steam reforming of ethanol into hydrogen-rich gas using microwave Ar/water "tornado"-Type plasma. Int J Hydrogen Energ. 2014; 39(11): 5663–5670.
[30] Kim DJ, Choi Y, Kim KS. Effects of process variables on NO_x conversion by pulsed corona discharge process. Plasma Chem Plasma P. 2001; 21(4): 625–650.
[31] Bian WJ, Song XH, Shi JW, Yin XL. Nitrogen fixed into HNO_3 by pulsed high voltage discharge. J Electrostat. 2012; 70(3): 317–326.
[32] Jolibois J, Takashima K, Mizuno A. Application of a non-thermal surface plasma discharge in wet condition for gas exhaust treatment: NO_x removal. J Electrostat. 2012; 70(3): 300–308.
[33] Shi JW. Study on nitrogen fixation into water by the technique of pulsed high-voltage discharge plasma and the mechanism. Suzhou University. 2010.
[34] Li K, Tang XL, Yi HH, Ning P, Ye ZQ, Kang DJ, Song JH. Non-thermal plasma assisted catalytic oxidation NO over Mn–Ni–O_x catalysts at low-temperature. Adv Mater Res-Switz. 2012; 383–390: 3092–3098.
[35] Nagayama E, Imura T, Nfizuno A. Plasma-catalytic combustion system for VOC removal. 25th ICPIG international conference on phenomena in ionized gases, 2001; 104: 109–110.
[36] Goujard V, Tatibouet JM, Batiot-Dupeyrat C. Influence of the plasma power supply nature on the plasma-catalyst synergism for the carbon dioxide reforming of methane. IEEE T Plasma Sci. 2009; 37(12): 2342–2346.

[37] Zhang YP, Zhu XL, Pan YX, Liu CJ. Improvement of coke resistance performance of Ni-Based catalysts in methane reforming via glow discharge plasma treatment. Chinese J Catal. 2008; 29(10): 1058–1066.

[38] Haryanto A, Fernando S, Murali N, Adhikari S. Current status of hydrogen production techniques by steam reforming of ethanol: A review. Energ Fuel. 2005; 19(5): 2098–2106.

[39] Zhang CX, Li SR, Wu GW, Huang ZQ, Han ZP, Wang T, Gong JL. Steam reforming of ethanol over skeletal Ni-based catalysts: A temperature programmed desorption and kinetic study. AIChE J. 2014; 60(2): 635–644.

[40] Comas J, Marino F, Laborde M, Amadeo N. Bio-ethanol steam reforming on Ni/Al_2O_3 catalyst. Chem Eng J. 2004; 98(1–2): 61–68.

[41] Batista MS, Santos RKS, Assaf EM, Assaf JM, Ticianelli EA. High efficiency steam reforming of ethanol by cobalt-based catalysts. J Power Sources. 2004; 134(1): 27–32.

[42] Wu YJ, Santos JC, Li P, Yu JG, Cunha AF, Rodrigues AE. Simplified kinetic model for steam reforming of ethanol on a Ni/Al_2O_3 catalyst. Can J Chem Eng. 2014; 92(1): 116–130.

[43] Fatsikostas AN, Verykios XE. Reaction network of steam reforming of ethanol over Ni-based catalysts. J Catal. 2004; 225(2): 439–452.

[44] Fierro V, Akdim O, Provendier H, Mirodatos C. Ethanol oxidative steam reforming over Ni-based catalysts. J Power Sources. 2005; 145(2): 659–666.

[45] Diagne C, Idriss H, Kiennemann A. Hydrogen production by ethanol reforming over Rh/CeO_2–ZrO_2 catalysts. Catal Commun, 2002; 3(12): 565–571(7).

[46] Roy B, Martinez U, Loganathan K, Datye AK, Leclerc CA. Effect of preparation methods on the performance of Ni/Al_2O_3 catalysts for aqueous-phase reforming of ethanol: Part I— catalytic activity. Int J Hydrogen Energ. 2012; 37(10): 8143–8153.

[47] Roland U, Holzer F, Kopinke ED. Combination of non-thermal plasma and heterogeneous catalysis for oxidation of volatile organic compounds Part 2. Ozone decomposition and deactivation of gamma-Al_2O_3. Appl Catal B-Environ. 2005; 58(3–4): 217–226.

[48] Rico VJ, Hueso JL, Cotrino J, Gallardo V, Sarmiento B, Brey JJ, González-Elipe AR. Hybrid catalytic-DBD plasma reactor for the production of hydrogen and preferential CO oxidation (CO-PROX) at reduced temperatures. Chem Commun. 2009; 41(41): 6192–6194.

[49] Krawczyk K, Mlotek M, Ulejczyk B, Schmidt-Szalowski K. Methane conversion with carbon dioxide in plasma-catalytic system. Fuel. 2014; 117(1): 608–617.

[50] El-Saftawy AA, Elfalaky A, Ragheb MS, Zakhary SG. Electron beam induced surface modifications of PET film. Radiat Phys Chem. 2014; 102: 96–102.

[51] Skacelova D, Stupavska M, St'ahel P, Cernak M. Modification of (111) and (100) silicon in atmospheric pressure plasma. Appl Surf Sci. 2014; 312(3): 203–7.

[52] Hou ZY, Yashima T. Meso-porous Ni/Mg/Al catalysts for methane reforming with CO_2. Appl Catal a-Gen. 2004; 261(2): 205–209.

[53] Rahemi N, Haghighi M, Babaluo AA, Jafari MF. Syngas production via CO_2 reforming of methane over plasma assisted synthesized Ni–Co/Al_2O_3–ZrO_2 nanocatalysts with different Ni-loadings. Int J Energ Res. 2014; 38(6): 765–779.

[54] Cheng DG, Zhu XL, Ben YH, He F, Cui L, Liu CJ. Carbon dioxide reforming of methane over Ni/Al_2O_3 treated with glow discharge plasma. Catal Today. 2006; 115(1): 205–210.

[55] Estifaee P, Haghighi M, Babaluo AA, Rahemi N, Jafari MF. The beneficial use of non-thermal plasma in synthesis of Ni/Al_2O_3–MgO nanocatalyst used in hydrogen production from reforming of CH_4/CO_2 greenhouse gases. J Power Sources. 2014; 257(3): 364–373.

[56] Liu GH, Li YL, Chu W, Shi XY, Dai XY, Yin YX. Plasma-assisted preparation of Ni/SiO_2 catalyst using atmospheric high frequency cold plasma jet. Catal Commun. 2008; 9(6): 1087–1091.

[57] Wang ZJ, Zhao Y, Cui L, Du H, Yao P, Liu CJ. CO_2 reforming of methane over argon plasma reduced Rh/Al_2O_3 catalyst: a case study of alternative catalyst reduction via non-hydrogen plasmas. Green Chem. 2007; 9(6): 554–559.

[58] Yu KL, Liu CJ, Zhang YP, He F, Zhu XL, Eliasson B. The preparation and characterization of highly dispersed PdO over alumina for low-temperature combustion of methane. Plasma Chem Plasma P. 2004; 24(3): 393–403.

[59] Zou JJ, Liu CJ, Zhang YP. Control of the metal-support interface of NiO-loaded photocatalysts via cold plasma treatment. Langmuir. 2006; 22(5): 2334–2339.

[60] Pribytkov AS, Baeva GN, Telegina NS, Tarasov AL, Stakheev AY, Tel'nov AV, Golubeva VN. Effect of electron irradiation on the catalytic properties of supported Pd catalysts. Kinet Catal. 2006; 47(5): 765–769.

[61] Jun J, Kim JC, Shin JH, Lee KW, Baek YS. Effect of electron beam irradiation on CO_2 reforming of methane over Ni/Al_2O_3 catalysts. Radiat Phys Chem. 2004; 71(6): 1095–1101.

[62] Rahemi N, Haghighi M, Babaluo AA, Jafari MF, Khorram S. Non-thermal plasma assisted synthesis and physicochemical characterizations of Co and Cu doped Ni/Al_2O_3 nanocatalysts used for dry reforming of methane. Int J Hydrogen Energ. 2013; 38(36): 16048–16061.

[63] Xu Z, Li YM, Zhang JY, Chang L, Zhou RQ, Duan ZT. Bound-state Ni species— A superior form in Ni–based catalyst for CH_4/CO_2 reforming. Appl Catal a-Gen, 2001; 210(1–2): 45–53.

[64] Sehested J, Gelten JAP, Remediakis IN, Bengaard H, Norskov JK. Sintering of nickel steam-reforming catalysts: Effects of temperature and steam and hydrogen pressures. J Catal. 2004; 223(2): 432–443.

[65] Sekine Y, Haraguchi M, Tomioka M, Matsukata M, Kikuchi E. Low-temperature hydrogen production by highly efficient catalytic system assisted by an electric field. J Phys Chem A. 2010; 114(11): 3824–3833.

[66] Skrzypek J, Sloczynski S, Ledakowicz S. Methanol synthesis. Warsaw: Polish

Scientific Publishers. 1994.
[67] Halonen L, Bernasek SL, Nesbitt DJ. Reactivity of vibrationally excited methane on nickel surfaces. J Chem Phys. 2001; 115(12): 5611–5619.
[68] Schmid MP, Maroni P, Beck RD, Rizzo TR. Molecular-beam/surface-science apparatus for state-resolved chemisorption studies using pulsed-laser preparation. Rev Sci Instrum. 2003; 74(9): 4110–4120.
[69] Beck RD, Maroni P, Papageorgopoulos DC, Dang TT, Schmid MP, Rizzo TR. Vibrational mode-specific reaction of methane on a nickel surface. Science. 2003; 302(5642): 98–100.
[70] Nishiguchi T, Matsumoto T, Kanai H, Utani K, Matsumura Y, Shen WJ, Imamura S. Catalytic steam reforming of ethanol to produce hydrogen and acetone. Appl Catal A-Gen. 2005; 279(1): 273–277.
[71] Aupretre F, Descorme C, Duprez D. Bio-ethanol catalytic steam reforming over supported metal catalysts. Catal Commun. 2002; 3(6): 263–267.
[72] Kim SS, Lee H, Na BK, Song HK. Plasma-assisted reduction of supported metal catalyst using atmospheric dielectric-barrier discharge. Catal Today. 2004; 89(1–2): 193–200.

Chapter 7
Outlook

In the field of ethanol reforming, both the non-thermal plasma reforming process and the conventional catalyst reforming process are the widespread focus of investigation. In some literature, it has been pointed out that non-thermal plasma combined with a catalyst can efficiently improve the reforming efficiency. However, the non-thermal plasma-catalyst coupled system is very complex in the reforming reaction, and there is little literature that focuses on the overall introduction of the synergistic effect in the coupled reformation. Furthermore, the plasma reaction, the catalyst surface reaction and the synergistic effect are involved in the reforming process and plasma can modify the catalyst surface; all these factors bring a great challenge to theoretical numerical simulation. Based on the development of the theory of plasma fuel reforming and material reformation, the practice of the catalyst reforming process and the gradually maturing theoretical level, however, investigation on the mechanism and the numerical simulation of non-thermal arc plasma should be developed further. In the application on vehicles and mobile power, the non-thermal plasma reaction is always characterized by a low reaction temperature, high reactivation, a short response time, a small setup size and high energy efficiency. The conventional catalyst process also has the advantages of high conversion efficiency and high reactivity orientation. Hence, the non-thermal catalyst is a promising process to be widely used in vehicle applications. Furthermore, the miniaturization, integration, and low energy efficiency of non-thermal arc technology make it possible to apply it in the field of portable power; at that time, it will be possible to apply the non-thermal plasma process to a very large area ranging from power plants, public traffic to mobile phones.

Furthermore, since non-thermal plasma is very promising in fuel reforming, in the preparation of nanomaterials, in the surface modification of materials, in medical sterilization and in the treatment of environmental protection, it can be predicted that normalized and universal non-thermal plasma technology can be realized, paving the way for a low-cost, diversified/multifunctional non-thermal plasma process.

Index

Alcohols	3, 17, 19
Catalyst	7, 51, 86
Ethanol	1, 6, 15, 31
Flow rate	9, 38, 46
Hydrogen	6, 15, 31
Mechanism	59, 81, 82
Non-thermal arc	15, 17, 25
NO_x	59, 80, 81
O/C ratio	34, 44, 53
Plasma	1, 8, 15
Plasma reforming system	31, 41, 93
Plasma-catalytic	3, 49, 81
Radical reaction	37, 59, 67
Reforming	1, 3, 4
S/C ratio	37, 45, 55